WHAT IS EARTH?

WHAT IS EARTH?

ABDULAZIZ ALNAZARI

Twitter: AbdulAziz_Mohd
Instagram: az_abdulaziz
GoodReads: AbdulAziz_Mohammed

ISBN-13: 9781707973941

UAE National Media Council Approval Number: MC-01-01-1946043
Age Classification: E

In the Name of Allah, the Most Merciful, the Most Gracious

DEDICATION

To those hiding their comfort and delight with the statement of Allah, Exalted and Glorified, **"And at the Earth—how it is spread out?"** [Al-Ghāshiyah:20].

May I remind you of a Quranic verse following Allah's, Exalted and Glorified, saying, **"So remind, [O Muhammad]; you are only a reminder. You are not over them a controller"** [Al-Ghāshiyah:21–22].

And this verse, **"And if you obey most of those upon the Earth, they will mislead you from the way of Allah. They follow not except assumption, and they are not but falsifying"** [Al-An'ām:116].

"And peace will be upon he who follows the guidance" [Tāhā:47].

TABLE OF CONTENTS

Dedication · vii

Introduction ·xi

1   Why Discuss the Shape of the Earth? · · · · · · · · · · · · · · · · 1

2   Quran and Earth · 6

3   Heavens and Earth · 9

4   Earth as a Bed · 21

5   Sky is a ceiling · 28

6   Earth is Spread out · 32

7   Seven Heavens · 39

8   Succession of the Creation of the Heavens and Earth · · · · · · · · · · 50

9   Beginning of the Heavens and Earth · · · · · · · · · · · · · · · · · · · 60

10 Six Days · 70

11 Width of the Heavens and Earth · 75

12 Fall of the Sky Upon the Earth · 81

13 Qibla (Prayer Direction) · 86

14 Allah Set Firm mountains · 89

15 Traveling Through the Earth · 94

16 Quick Reflections on the Quran · · · · · · · · · · · · · · · · · · · 100

17 Refutation of the Allegedly Quran-supported Claims that
the Earth is Spherical ·139

18 Refutation of Sheikh Muhammad Al-Amin Al-Shinqiti of
the argumentation of some contemporaries · · · · · · · · · · · · · · · 166

19 Prophetic Hadiths and Flat earth Theory · · · · · · · · · · · · · · · ·178

20 Bible and Flat Earth Theory · 180
21 Religions, Mythologies and the Flat Earth Theory · · · · · · · · · · · · 202
22 Questions and General Proofs · 203
23 Night and Day · 224
24 The Sun · 235
25 The Moon ·247
26 Solar and Lunar Eclipses · 254
27 Deluge of Noah · 258
28 Gravity · 262
29 Satellites · 271
30 Moon Landing · 275
31 Images of the Earth · 279
32 Photo of a Black Hole · 282
33 Conspiracy · 285

Postscript · 295
Appendix 1: Preference for the Truth over the Creation · · · · · · · · 297
Appendix 2: Refutation of the Sermon of Sheikh Adnan Ibrahim:
            Agents of NASA and the Sphericalness of the Earth · · · 307

INTRODUCTION

S INCE CHILDHOOD, I have always found myself in the middle of discussions and debates with people about diverse topics, covering issues of belief and Sharia, philosophy and intellect, sociology, and even entertainment. At the end of these dialogues, I would come out either as a teacher or as a learner—happy or shocked—because the people around me would not approve many of the ideas I would discuss with them. Occasionally, I would be shocked at the ideas proposed by others. One of the topics we would discuss in our debates, whether in the real world or the virtual one through the World Wide Web, was the shape of the Earth: Is it spherical or flat?

This question has occupied the minds of people for more than a thousand years. The question may seem strange—even asking it may sound ridiculous for some people in our current time—but it is not so for many who wield knowledge. At the time, I was like all the other enlightened copycats who believe themselves free from the shackles of imitation but are actually imitators of a class or a community of people than individuals. I was as impressed with the achievements of modern science as the others. I would be fascinated by the successive scientific discoveries, with a special interest in cosmology, reading many introductory and specialized books on this discipline, in addition to watching countless movies and documentaries and listening to hundreds of hours of podcasts.

In my bubble, I thought of myself as an expert on cosmology in my society, but I spoke very little on the subject, given the lukewarm interest of people around me, including those with whom I chatted on the social media platforms of the time. I found no serious interest that would force me to highlight the subject.

Frankly speaking, most of the scientific books I read and series I watched were enough to arouse doubts at the heart of a Muslim regarding the Quranic view of the shape of the Earth. However, before venturing into the midst of modern science, I was armed with the knowledge I gained from many books that specialized on the scientific inimitability of the Quran and the compelling arguments the early scholars, such as ibn Taymiyyah, ibn al-Qayyim, Abu Hāmid al-Ghazāli, al-Fakhr al-Rāzi, and al-Qādi Abd al-Jabbār al-Hamadhāni (may Allah have mercy upon them) proposed in support of the belief in the existence of Allah, His angels, scriptures, messengers, and the Last Day.

Anytime I came across arguments advocating atheism, visibly or in disguise, I would refute those arguments by more logical and rational arguments. I am indebted to all our scholars, despite their various ideologies or specific affiliations, for what I have learned from them; may Allah have mercy upon them all and make them among the inheritors of the Gardens of Pleasure. Allah says, **"And whoever obeys Allah and the Messenger—those will be with the ones upon whom Allah has bestowed favor of the prophets, the steadfast affirmers of truth, the martyrs and the righteous. And excellent are those as companions"** [Al-Nisā':69].

On many occasions, I would quote the arguments of the scientific inimitability of the Quran, as offered by scholars and contemporary intellectuals like Sheikh al-Sha'raawi, Dr. Mustafa Mahmoud, and Dr. Zaghlūl al-Najjār. I would speak with my siblings about the magnificence of the Quran and how it is a miraculous Book, considering the scientific facts stated long before they were discovered by scientists of today.

Still, I must be frank with myself and my dear readers. All this visual, audible, and written material was not enough to answer the questions I had about the shape of the Earth. For this reason, I used to occasionally provoke my younger brother or some of my friends with the question, what if Earth is flat? After in-depth research into Quranic studies and Sharia sciences in general, I developed a stronger belief in the similarity between the Quranic outlook to the shape of Earth and that of the holy scriptures of the Jews and Christians. I went further with my research and reached to some conclusions advocating a different shape of the Earth than propagated by the media and maintained by the contemporary scientific communities.

Eventually, after the topic gained popularity among Muslims and Arabs, I found many of the writers, Muslim scientists, and Arab intellectuals ridiculing those who believe the Earth is flat in satiric and provocative articles devoid of knowledge that sparked me to write about this subject. Considered to be respected intellectuals, they were not only ridiculing the opposing views but also perpetrating various poor practices they themselves condemn in their books and articles.

For example, they accuse their opponents who believe the Earth is flat of imitation and backwardness. They refer to imitation of some of the Salafi sheikhs such as Sheikh ibn Bāz or Sheikh ibn ʿUthaymīn (may Allah have mercy upon them), despite the fact that their fatwas addressed the motion of the Earth and the sun without any details regarding the shape of the Earth, let alone whether it is flat or spherical.

However, many of these intellectuals ridiculing lack the essential knowledge of the subject except for few bits and pieces, and they are the ones who imitate the scientific community and the unstoppable media propaganda, as if they are broadcasting an active revelation from NASA and other space agencies. They also accuse others with the exercise of intellectual terrorism and radicalism while approving it for themselves. They intellectually terrorize those believing the Earth is flat by saying, "The whole word is unanimous against you, and you are merely a small group soon to disappear like salt dissolving in water." They may even say, "The 'x' government or the 'x' superpower is planning to build a community in Mars, invade the space, or send a spacecraft to the 'x' planet. The West has made it to the moon, while you are still arguing about the shape of the Earth, flat or spherical."

In this scenario, the victim is the person who has a positive curiosity in searching for the truth. Once he hears or reads the preceding statement, he withholds researching those issues and ends up following those writers to avoid being an outcast in the society or the object of ridicule by his kin, and people in general, for no other reason but the intellectual terrorism exercised by those claiming tolerance, coexistence, and acceptance of others.

Hence, I decided to write down some of my thoughts, hoping it may spark people's curiosity to look for the truth. Any serious person would research this issue personally and break free from the radical shackles of imitation holding them back. I delayed writing the book for some of the reasons I mentioned here, in addition to my belief that I would not be offering any new insights for the Arab reader. However, after I read the articles of the intellectuals and thinkers and heard the statements of the contemporary scientists, I started thinking that I can benefit the people, even if only a little. My conscience made me feel guilty for refraining from writing on this subject while leaving the room for imitators. I began to write the book hoping my readers find it original and authentic.

My purpose is to show Arab Muslims how maintaining the Earth is flat is not backwardness, nor is it baseless or based on shaky proofs. Instead, it is an opinion supported by recognized evidence valued by reasonable people. At any case, I do not follow any single school of thought or any single doctrine, nor do I like to be molded into a specific ideology other than the general framework of Islam. In other words, I'm not a Sufi, Salafi, Mu'tazily, or Ash'ary. Thanks to Allah for being unbiased toward any group, I am unfazed by any group name.

I am aware this book will expose me to all sorts of criticism and may even affect my livelihood. It is even possible that my family may be caught in the crosshairs. However, it is no more than a temporal harm, which is trivial to begin with. I hope I become one of those who entrust their affairs with Allah and flee to Him, Almighty.

This book may be slightly burdensome for readers disinterested in Quranic studies or Tafsīr because I have extensively quoted the opinions of early Muslim scholars for various reasons, most importantly to relay their opinions regarding the shape of the Earth or to refute them, given that some Muslims today claim that scholars unanimously agreed that the Earth is spherical.

Since the book includes many uncommon opinions and new ideas—perhaps never thought of by the reader—that run counter to the widespread ideas, the reader may find it complicated and slightly hard to swallow, though nothing could be further from the truth. Because we have accustomed ourselves to easy and unprovocative material, I believe any book incapable of awakening, provoking, and challenging our mind-set without offering it new insights holds little benefit. Eventually, everything rests at the command of Allah.

The book is extensive to some extent, because excavating the truth of these subjects is not suited for lazy readers, nor is providing suggestive depictions enough. The reader of this book must be sincere and patient. Therefore, seek the help of Allah, Exalted and Glorified, Who is the only One to provide them. My advice to the reader is to hold a pen and a notebook, or even a smartphone, in order to draw and imagine the images I am communicating in this book, as that will help the reader understand the content of this book.

This book contains several chapters, and every significant issue connected to our topic has been discussed separately and extensively. I chose to ignore citing the volume and page numbers of sources pertaining the statements of the exegetes, because they are available and copiable through the internet, not to mention easy to find through search engines. I have also annexed two beneficial articles I wrote at an earlier time about the shape of the Earth. I named the first article "Preference for Truth over the Issue of Creation" and the second article "Refutation of the Sermon of Sheikh Adnan Ibrahim: Agents of NASA and the Sphericalness of the Earth."

In this book, I have enclosed a brief presentation of some of my views and opinions about the shape of the Earth and other relevant issues. I invite you, the reader, to read this book with an intent to search for the truth, not merely to refute the author of this book and his ideas. So, let us begin with seeking refuge in Allah and His assistance. I seek refuge in Allah from the cursed devil. In the name of Allah, the Most compassionate, the Most Merciful.

—AbdulAziz AlNazari

1

WHY DISCUSS THE SHAPE OF THE EARTH?

A T MANY OCCASIONS of discussing the shape of the Earth when the opponents have no valid argument, they would say, "What is the point of the arguments about this issue? It is very futile. It would not draw you closer to Allah and may even distance you owing to the argument."

True! This line of thinking is natural for some people. I, however, maintain this to not necessarily be true because studying the differences among scholars on this issue has many advantages. For one, it would not be the focal point of hundreds of verses Allah stated in detail about the Earth and heaven, had it not been significantly crucial. Even if it means nothing to you, it definitely meant a lot for Muslims interested in reflecting the Book of their Lord.

I find it strange to hear people questioning the benefit of this topic. Regardless of whether the Earth is spherical, flat, square, or even triangular, we worship Allah anyway and delving into such an issue is pointless. Fine, I will follow their line of argument. Assuming this is the case, why do not they argue the same thing about the many topics people talk about? There are thousands of popular topics, many of which arouse arguments and conflict, but they question the point of discussing such issues, although they are discussed by people of all affiliations who quote textual evidence from the Quran, Sunnah, and other scholarly statements. Why would they not ask those people the same question they ask those interested in the shape of the Earth? What is the point of learning about such-and-such thing as long as we worship Allah, the Most compassionate, the Most Merciful?

Try once to change between channels on TV to get an impression of the topics people discuss and argue about. You will find them countless. Why do not they question the benefit of those? Even if this topic of Earth is unimportant, why did you bother to ask about the significance of studying this topic? Dear reader, don't you think, as I do, that this type of people provoke interest in writing psychological, sociological, and philosophical studies?

Researching the shape of the Earth opens new horizons. The first thing you discover is that not everything you have learned in the past years is not necessarily true. Perhaps you may discover you have been taught lies, one after the other, without the slightest inkling or even a minor test. Once you discover you have been betrayed as a human being who deserves no deception, you may be driven to break free from the disease of imitation that is an evil plaguing people's lives and religion.

Dear reader, you should know that a Muslim is required to reflect on the verses of the Quran instead of glossing over or memorizing them without understanding. Certainly, the comprehension of Quranic verses plays an essential role in forming the belief of the individual and shaping his life.

One of the reasons studying the shape of the Earth is important is to identify the necessary steps to take when a Muslim finds Quranic verses showing the Earth is flat whereas the scientific community is broadcasting, day and night, that it is spherical. Either he loses trust in the Book of his Lord or believes it on its entirety without understanding the necessities of Allah's verses or the conclusions of the modern science. Put it simply, he understands neither of them. The other choice is to doubt the scientific theories and start researching. If he did so but ended up discovering that the sphericalness of the Earth is no more than a theory, what would happen?

The faith of the believer in the Book of Allah and the distinct signs about the shape of the Earth would increase and rehabilitate the image he developed about the prophetic reports proving the flatness of the Earth. Furthermore, he would acquire a new outlook on his living space and on the shape of the existence hosting the story of mankind. His belief in the messengers and scriptures will heighten as he notices the agreement between the Quran and the Bible on the issue of the shape of the Earth.

Not only will Muslims benefit from this but also Jews and Christians. Researching this issue and arriving at conclusions proving the invalidity of the sphericalness of the Earth combined with locating its flatness at the Bible— the Old and New Testaments—will rehabilitate the image about the Bible and induce its guidance. More in-depth research in other religions, in addition to the literal implications of the Quranic verses proving the flatness of the earth, will increases the chances of imparting the belief of Allah and embracing the broad religion of Islam by the Jewish or the Christian out of adherence rather than imitation. These chances grow slimmer if the issue of researching the shape of the Earth goes uninvestigated.

This issue will definitely leave an impact on disbelievers in Allah or atheists who believe the inexistence of a God. Many atheists believe in the conclusions of the modern science such as the Big Bang that created the cosmos, resulting in the expansion of the cosmos and the creation of galaxies, stars, planets, suns, and even the Earth. Afterward, life began and evolved to its final stage represented in the human being. The Earth is no more than an atom in a vast galaxy like a ring in a huge desert. This galaxy, the Milky Way, is in itself an atom compared to the vastness of this cosmos. The great design of this cosmos is merely an accident caused by a chaotic explosion in the cosmos.

However, when the atheist is shocked after learning most of what he was told is untrue, he will learn he is living in a space like his home, with a floor to sleep on, a sky for shade like a house ceiling, surrounded by impenetrable walls without any doors to escape except after death. He may ask himself, "Who placed me in this house? Why am I here? What does He want from me?" Questions like these, dear reader, make all the difference in human souls. It has done for thousands of years and still retains its effect today, and all praise is due to Allah, the Lord of the worlds.

Today, if you ask some scholars of Islamic studies and preachers if the Earth is spherical or flat in the Quran, they will be angered by your question and talk loudly as if you offended them. But why such anger? Consider the questioner a seeker of a fatwa. Do they usually treat questioners with such anger even if their questions seem strange? Of course, they do not. So, why do you get angry when a confused Muslim asks an innocent question: is the Earth flat or spherical? The same reaction is produced by thinkers and intellectuals who are raged by the mere question. They stab the questioner with their verbal assault. Is it not this a proof of a deep belief and a strange love for the conclusions made by modern science regarding the shape of the Earth? Is their rage caused by the fear of destabilizing their belief? Or have some of them become similar to those described in Allah's saying, **"And their hearts absorbed [the worship of] the calf"** [Al-Baqarah:93].

This issue need not provoke any anger. Let me give a word to the peaceful who think well of people; thanks to Allah, they are good people. When some Muslim investigates the issue of the shape of the Earth and discovers the many mistakes of the scientists of today, the halo of sanctifying scholars will disappear, and people will come to the realization that the scientists are the same as other people who err. Subsequently, it will be easier for people not to agree with them at everything they say. They will discover the presence of many sources channeling lies and misconceptions for hundreds of years, either due to imitation or purposefully. If it is the former, they will hate imitation, but if is the latter, they will discover they were deceived intentionally by media outlets and realize they are living in a huge lie. They will take precautions to avoid being the victims of those lies and begin educating the public, starting from their families to all people.

Millions have been spent to spread the idea of the sphericalness of the Earth, and a great number of scientists were slandered and psychologically traumatized for no other reason but for their contradicting the perspective of the sphericalness of the Earth. Countless number of people who were deceived by this lie or mistake deserve to be told the truth. People have lost their free choice at the expense of echoing the sphericalness of the Earth. Aren't these human casualties enough? Doesn't this require us to pause for a moment and restart studying the issue of the shape of the Earth from scratch? The importance of studying this issue is significant at many levels.

In brief, studying this issue will help us recognize truth from falsehood and right from wrong. The human being would not be where he is today without paying due importance to such issues. Allah said, "Say, **'Is the blind equivalent to the seeing? Or is darkness equivalent to light?'**" [Al-Ra'd:16]

2

QURAN AND EARTH

I N THIS CHAPTER, I will show the reader some references to the shape of the Earth in the Quran. In this chapter and the following one, I have included some of my comments and views about the Quranic verses. Perhaps they may help you understand that this topic is not as easy as many people believe. In addition to this, following chapters contain some commentaries of exegetes on those verses. Let us reflect on them since they are relevant to our topic. Before I begin, I would like to highlight some important points.

Anyone investigating the Quran will find no explicit verse or a part of a verse indicating the Earth is spherical or a description in that near vicinity. By contrast, there are multiple statements supporting the idea it is flat, such as **"[He] who made for you the earth a bed [spread out]"** [Al-Baqarah:22]; **"Have We not made the earth as a bed?"** [Al-Naba':6]; **"And Allah has made for you the earth an expanse"** [Nūh:19]; **"And at the earth—how it is spread out?"** [Al-Ghāshiyah:20]; **"And when the earth has been extended"** [Al-Inshiqāq:3]; **"And [by] the earth and He who spread it"** [Al-Shams:6]; **"And you see the earth barren"** [Al-Hajj:5]; and, **"And of His signs is that you see the earth stilled"** [Fuṣṣilat:39].

There are other words [in the Quran] indicating the Earth is flat, stationary, but revolving. If one argues that words in the Quran are meant to suit people's reason and that people cannot bear the idea that the Earth is not flat, this is a weak argument. Allah, Exalted and Glorified, obligated in His Book the belief in greater issues than the one in question. In addition, this would require the proof of the sphericalness of the Earth must be initiated by those claiming the Earth is spherical, not by those believing it is flat.

All the available evidence indicating the Earth is spherical are susceptible to logical refutation undriven by stubbornness or ridiculousness. Though some of the modernists do not slander those contradicting the Quran and Sunnah as long as the contested evidence allows for a valid interpretation, but, ironically, they slander those believing the Earth is flat though most of the logical and textual evidence supports the latter's rationale. Today, even visual and experimental evidence are available for both parties, but I am reluctant to rely on them because both parties may contradict each other and accuse each other's intentions.

I would like to say that The Quran, from the initial chapter of al-Fātihah until al-Nās, proves the Earth is flat. Before commenting on some of the Quranic verses, let's play a game! What do you think? Okay, then. Look at this puzzle. A frame with a space of 3×3 from wooden pieces, written on top of them the following numbers in this order:

$$2 \quad 9 \quad 4$$
$$7 \quad 5 \quad 3$$
$$6 \quad 1 \quad 8$$

In this puzzle, the addition of each column or row from any direction equals 15. The addition of the first row, for example, adds up to 15. The same 15 is the outcome of the left column composed of 2, 7, and 6. Also, if we do it diagonally in either directions, the addition of numbers 4, 5, and 6 on the first set, and numbers 2, 5, and 8 on the second set equal 15.

Try to replace number 5 with any number from 1 to 9 except for 5 to produce the outcome 15. What would happen? You will not succeed, won't you? Before replacing number 5, there were 8 ways to conjure up number 15, whereas after replacement, the ways to produce 15 decreased to only four ways, which is half the previous number. Did you notice the effect of replacing a number in the puzzle? It has dramatically disrupted the great design, cutting it to half of what it is.

The same applies to those who attempts to twist the Quranic verses and deviate their interpretation beyond their literal sense to suit their desires. The Quran is a glorious Book laying out an originally perfect system. Should we take it seriously as we were given it, without changing the meanings of verses intentionally and unintentionally, we will discover many readily understandable issues. In contrast, if our interpretation deviates to correspond with the conclusions of modern science, we will be in a dilemma. It is possible to deviate the interpretation of a given verse just to prove the sphericalness of the Earth, but how about the rest of the verses? How would a regular layman who notices the flatness of the earth from some Quranic verses and its sphericalness from other verses react? How would he react upon seeing the language of the Quranic verses losing its literal and lexical meaning? Would he not perceive a misdirection to prove the Quran is in correspondence with the conclusions of modern science?

For this reason, dear reader, even if some attempted to prove that some Quranic verses indicate the sphericalness of the Earth, tell them to pause for a while and read the entire Quranic text to understand the holistic view this Quran offers about this existence. You will come to discover the error in the particular instance they interpreted in a way to be consistent with modern science. Some contemporaries have, unfortunately, done the same as replacing number 5 in the puzzle, thereby disturbing parts of the system. Allah, Almighty, said, "**And among them are unlettered ones who do not know the Scripture except in wishful thinking, but they are only assuming. So, woe to those who write the 'scripture' with their own hands, then say, 'This is from Allah,' in order to exchange it for a small price. Woe to them for what their hands have written and woe to them for what they earn**" [Al-Baqarah:78–79].

3

HEAVENS AND EARTH

THIS CHAPTER IS fairly complex, and it may be hard to comprehend its under-lying objectives, especially if skimmed. It will be easier, however, if read deliberately. This chapter introduces the main topics I will discuss elaborately in the following chapters, and the next chapters complement and elucidate this one. If you wish, you can skip this chapter and return to it after you finish read-ing. If you also wish, you can read it deliberately so that you can discover the true depth of what I will lay out in the following chapters. Should you wish to leave no stone unturned and grasp my far-reaching and profound implications, try to read the book more than once, and you shall find treasures therein.

It seems you have decided to continue reading. Excellent! Let us begin by posing a question: "What are the heavens and the Earth?" It is a fact demanding you to think of it considerably before continuing to read. What are the reasons Allah referred to the heavens and the Earth in numerous instances? I believe the heavens and the Earth represent the cosmos in our current terminology. It is the place that hosts the human drama in this life. For me, there is a difference between the "heavens and the Earth," "the seven heavens," "sky," "atmosphere of the sky," and "the Earth." I will elaborate on some of these points briefly, because the topic is complex and requires an accurate understanding.

You may ask, "What is your proof that the heavens and the Earth refer to the cosmos or that they contain everything, but not necessarily all that exists?" To know the answer, you must keep reading. When both "the heavens and the Earth" are coupled in the Quran, the first thing that comes to my mind is the cosmos in current terminology. When it is only "the seven heavens" without the Earth, it refers to all that is above the earth, including the celestial spheres, the pathways, or the planets orbiting in their respective pathways. When it is only the "sky," it may indicate only the base meaning of highness: the lowest, first, or the separating sky, or the ceiling carrying creatures on top of it, or some of the divine commands, such as rain, punishment, and revelation.

When the Quran only mentions "the atmosphere of the sky," it refers to the space between the first (or lowest) sky and the earth, or the separation between the ceiling sky and the earth, hosting birds, planes, clouds, and so on. When it is only "the earth," it refers to a part of the earth, a piece of land, or the ground that represents the underside of the presence of the cosmos. Quickly, let me explore some of the verses to clarify the subtle distinctions, for the Quran is the clear Book and the lantern that alleviates our difficulties of understanding. In addition, the Quran explains itself by itself.

THE HEAVENS AND THE EARTH

When Allah, Almighty, speaks about existence or the cosmos (by current definition), He always couples the heavens and the Earth. He does not mention them separately to show the reader that the correct name of the cosmos is the heavens and the Earth rather than just the cosmos. The so-called cosmos is composed of two parts: the topside and underside. Allah, Almighty, said, "**It is He who created the heavens and earth in six days and then established Himself above the Throne**" [Al-ḥadīd:4].

There is another great verse clearly demonstrating the distinction. It says, "**Indeed, in the creation of the heavens and earth, and the alternation of the night and the day, and the [great] ships which sail through the sea with that which benefits people, and what Allah has sent down from the heaven of rain, giving life thereby to the earth after its lifelessness and dispersing therein every [kind of] moving creature, and [His] directing of the winds and the clouds controlled between the heaven and the earth are signs for a people who use reason**" [al-Baqarah:164].

If you reflect, you will notice the verse is initiated by stating "the heavens and the earth," which refers to the existence I named the cosmos, including its topside and underside. In other words, it refers to the complete Cosmic Bubble or the Cosmic Egg. Next is **"and what Allah has sent down from the heaven of rain."** This refers to the sky above us, which is the topside of existence. Then, He, Almighty, mentioned "the earth," which represents the underside of existence. Lastly, He, Almighty, mentioned, **"and the clouds controlled between the heaven and the earth,"** which refers to the atmosphere of the sky that lies between the lowest sky and the earth. Read the verse again, and you shall find this is an accurate understanding of the verse. It is more precise than considering them to share identical meaning.

In the chapter of Ibrāhīm, Allah, Almighty, mentioned initially the creation of existence followed by the creation of the lowest sky. I hope you read this very carefully.

"It is Allah who created the heavens and the earth and sent down rain from the sky and produced thereby some fruits as provision for you and subjected for you the ships to sail through the sea by His command and subjected for you the rivers" [Ibrāhīm:32]. In the chapter of al-Naml, **"[More precisely], is He [not best] who created the heavens and the earth and sent down for you rain from the sky, causing to grow thereby gardens of joyful beauty which you could not [otherwise] have grown the trees thereof? Is there a deity with Allah? [No], but they are a people who ascribe equals [to Him]"** [al-Naml:60].

Consider more verses that explicitly refer to existence, currently called the "cosmos," as "the heavens and the Earth." Read, **"And to Allah belong the unseen [aspects] of the heavens and the earth and to Him will be returned the matter, all of it, so worship Him and rely upon Him. And your Lord is not unaware of that which you do"** [Hūd:123].

And, **"Say, 'Who is Lord of the heavens and earth?' Say, 'Allah.' Say, 'Have you then taken besides Him allies not possessing [even] for themselves any benefit or any harm?' Say, 'Is the blind equivalent to the seeing? Or is darkness equivalent to light? Or have they attributed to Allah partners who created like His creation so that the** creation [of each] seemed similar to them?' Say, 'Allah is the Creator of all things, and He is the One, the Prevailing" [al-Raʿd:16].

And, "Allah is the One who created the heavens and the Earth" [Ibrāhīm:32].

And, "[Moses] said, 'You have already known that none has sent down these [signs] except the Lord of the heavens and the earth as evidence, and indeed I think, O Pharaoh, that you are destroyed'" [Al-Isrā':102].

And, "Allah created the heavens and the earth in truth. Indeed in that is a sign for the believers" [Al-ʿankabūt:44].

And, "And if you asked them, 'Who created the heavens and the earth?' they would surely say, 'Allah.' Say, 'Then have you considered what you invoke besides Allah? If Allah intended me harm, are they removers of His harm; or if He intended me mercy, are they withholders of His mercy?' Say, 'Sufficient for me is Allah; upon Him [alone] rely the [wise] reliers'" [Al-Zumar:38].

And, "Indeed, Allah knows the unseen [aspects] of the heavens and the earth. And Allah is Seeing of what you do" [Al-ḥujurāt:14].

And, "And We made firm their hearts when they stood up and said, 'Our Lord is the Lord of the heavens and the earth. Never will we invoke besides Him any deity. We would have certainly spoken, then, an excessive transgression'" [Al-kahf:14].

And, "To whom belongs the dominion of the heavens and the earth. And Allah, over all things, is Witness" [Al-burūj:9].

And, "And to Him belongs whoever is in the heavens and earth. All are to Him devoutly obedient" [Al-Rūm:26].

And, "And to Allah belong the soldiers of the heavens and the earth. And ever is Allah Exalted in Might and Wise" [Al-Fatḥ:7].

THE SKY

As I mentioned, the sky refers to the lowest sky or the ceiling above us, below the high heavens. Allah said about it, "[He] who made for you the earth a bed [spread out] and the sky a ceiling" [Al-Baqarah:22]; and, "**And We made the sky a protected ceiling, but they, from its signs, are turning away**" [Al-Anbiyāa:32]; and, "**Have they not looked at the heaven above them—how We structured it and adorned it and [how] it has no rifts?**" [Qāf:6]; and, "**And when the heaven is split open and becomes rose-colored like oil**" [Al-Rahmān:37]; and, "**And the sky will split [open], for that Day it is infirm**" [Al-ḥāqah:16]; and, "**And [mention] the Day when the sky will split open with [emerging] clouds, and the angels will be sent down in successive descent**" [Al-Fur'qān:25]; and, "**And when the sky is opened**" [Al-Mur'salāt:9]; and, "**Are you a more difficult creation or is the sky? Allah constructed it**" [Al-Nāziʿāt:27]; and, "**When the sky breaks apart**" [Al-Infiṭār:1]; and, "**When the sky has split [open]**" [Al-Inshiqāq:1]. These verses do not address the high heavens but only the ceiling sky above us.

Allah, Exalted and Glorified, said, "**Do you not see that Allah has subjected to you whatever is on the earth and the ships which run through the sea by His command? And He restrains the sky from falling upon the earth, unless by His permission. Indeed Allah, to the people, is Kind and Merciful**" [Al-Hajj:65]. The "sky" mentioned here is the ceiling sky just above the atmosphere of the sky. The "earth" mentioned here does not refer to the earth in its entirety or everything on the underside of the cosmos but instead the land, and the clue being its mention in opposition to the sea.

This ceiling sky has gates, as mentioned in the saying of Allah, Exalted and Glorified, "**And [even] if We opened to them a gate from the sky and they continued therein to ascend**" [Al-ḥijr:14]; and, "**Then We opened the gates of the sky with rain pouring down**" [Al-Qamr:11]; and, "**And the sky is opened and will become gateways**" [Al-Naba':19]. Naturally, every construction has gates, as opposed to the space. It is inconceivable to have gates in the space. Those gates may as well be present on the high heavens. There are many hadith traditions that prove the presence of gates in the sky. However, whenever the word "gates" is mentioned in the Quran, it is related to the ceiling sky.

In a different sense, heaven refers to the source of the divine commands and the actions of God such as blessings, punishment, and revelation. If you wish, read the following verse, **"But those who wronged changed [those words] to a statement other than that which had been said to them, so We sent down upon those who wronged a punishment from the heaven because they were defiantly disobeying"** [Al-Baqarah:59]. In another verse, **"The People of the Scripture ask you to bring down to them a book from the heaven. But they had asked of Moses [even] greater than that and said, 'Show us Allah outright,' so the thunderbolt struck them for their wrongdoing. Then they took the calf [for worship] after clear evidences had come to them, and We pardoned that. And We gave Moses a clear authority"** [Al-Nisā':13]. Another verse states, **"And [remember] when they said, "O Allah, if this should be the truth from You, then rain down upon us stones from the sky or bring us a painful punishment"** [Al-Anfāl:32]; and, **"And O my people, ask forgiveness of your Lord and then repent to Him. He will send [rain from] the sky upon you in showers and increase you in strength [added] to your strength. And do not turn away, [being] criminals"** [Hūd:52]; and, **"It is He who shows you His signs and sends down to you from the sky, provision. But none will remember except he who turns back [in repentance]"** [Ghāfir:13]; and, **"If We willed, We could send down to them from the sky a sign for which their necks would remain humbled"** [Al-Shu'arā:4]; and, **"He will send [rain from] the sky upon you in [continuing] showers"** [Nūḥ:11].

THE ATMOSPHERE OF THE SKY

On this meaning, read the saying of Allah, Exalted and Glorified, **"Do they not see the birds controlled in the atmosphere of the sky? None holds them up except Allah. Indeed in that are signs for a people who believe"** [Al-Naḥl:79]; and, **"Do they not see the birds above them with wings outspread and [sometimes] folded in? None holds them**

[aloft] except the Most Merciful. Indeed He is, of all things, Seeing" [Al-Mulk:19].

This atmosphere is the space in which the human can fly and do anything he wants. It neither refers to the ceiling sky nor the heavens above the ceiling. Planes, rockets, and satellites are assumingly located on this atmosphere, and the same is true for the clouds. Read carefully the saying of Allah, Exalted and Glorified, in the chapter of al-Baqarah, **"and [His] directing of the winds and the clouds controlled between the sky and the earth are signs for a people who use reason."** He mentioned it is between the sky and the Earth. This space located in the middle is the atmosphere of the sky. It is the space where humans can penetrate and fly within. For this reason, birds do not fly on the seven heavens, and it is also the reason Allah has not said the subjected clouds between the heavens and the earth. Instead, Allah said, **"between the sky and the earth"** so that the human recognizes the difference.

As for the highness in general, Allah, Exalted and Glorified, said, **"Have you not considered how Allah presents an example, [making] a good word like a good tree, whose root is firmly fixed and its branches [high] in the sky?"** [Ib'rāhīm:24]; and, **"Whoever should think that Allah will not support [Prophet Muhammad] in this world and the Hereafter—let him extend a rope to the ceiling, then cut off [his breath], and let him see: will his effort remove that which enrages [him]?"** [Al-Hajj:15]; and, **"Do you feel secure that He who [holds authority] in the heaven would not cause the earth to swallow you and suddenly it would sway? Or do you feel secure that He who [holds authority] in the heaven would not send against you a storm of stones? Then you would know how [severe] was My warning"** [Al-Mulk:16–17].

THE SEVEN HEAVENS

This term has several meanings: pathways, high heavens, planets, and so on. Read the saying of Allah, Exalted and Glorified, **"It is He who created for you all of that which is on the earth. Then He directed Himself to** the heaven, [His being above all creation], and made them seven heavens, and He is Knowing of all things" [Al-Baqarah:29]; and, "He created the heavens without pillars that you see and has cast into the earth firmly set mountains, lest it should shift with you" [luqmān:10]. In this context, the word "heavens" does not refer to the entire cosmos because it is not coupled with the earth. Also, it is not the lowest sky because it is mentioned in the plural "heavens." In this verse, heavens refer to the high spheres, high heavens, or the high pathways. We shall discuss the seven heavens on the following chapters. The objective here is to promptly list some of the verses that show the difference.

"And He completed them as seven heavens within two days and inspired in each heaven its command. And We adorned the nearest heaven with lamps and as protection. That is the determination of the Exalted in Might, the Knowing" [Fuṣṣilat:12]; and, "And] who created seven heavens in layers. You do not see in the creation of the Most Merciful any inconsistency. So return [your] vision [to the sky]; do you see any breaks?" [Al-Mulk:3].

And, "They have not appraised Allah with true appraisal, while the earth entirely will be [within] His grip on the Day of Resurrection, and the heavens will be folded in His right hand. Exalted is He and high above what they associate with Him" [Al-Zumar:67]. The earth here represents the underside, while the heavens represent the topside. In the same sense, "Do you not see that to Allah prostrates whoever is in the heavens and whoever is on the earth and the sun, the moon, the stars, the mountains, the trees, the moving creatures and many of the people? But upon many the punishment has been justified. And he whom Allah humiliates—for him there is no bestower of honor. Indeed, Allah does what He wills" [Al-Hajj:18].

You may say that assuming my observation about the heavens and the earth to mean the cosmos in our modern language is true, why is the phrase "what is in between" coupled with them in many Quranic verses? And what is the difference between the structures?

It is true. Allah, Exalted and Glorified, mentions the phrase "what is in between" in many verses, but the underlying rationale is beyond the grasp of human reason. We are only intruding on the spacious fount of rationales instilled in the glorified Book of Allah. Allah, Exalted and Glorified, said, "**Do they not contemplate within themselves? Allah has not created the heavens and the earth and what is between them except in truth and for a specified term. And indeed, many of the people, in [the matter of] the meeting with their Lord, are disbelievers**" [Al-Rūm:8]; and, "**[Moses] said, 'The Lord of the heavens and earth and that between them, if you should be convinced'**" [Al-Shuʿarā:24]; and, "**Lord of the heavens and the earth and whatever is between them, the Exalted in Might, the Perpetual Forgiver**" [ṣād:66]; and, "**[From] the Lord of the heavens and the earth and whatever is between them, the Most Merciful. They possess not from Him [authority for] speech**" [Al-Nabaʾ:37], among many others.

In truth, every verse requires a rhetorical study and a genuine attempt to uncover its secrets. After deep analysis, it seems to me that the heavens and the earth are equal whenever Allah says, "Lord of the heavens and the earth." Accordingly, it is valid to say, "what is in between them." How is this possible given that the earth is a drop in the oceans when measured against the heavens? However, the phrase "in between them" indicates that the topside is equal, or at least approximate, in weight, to the underside. Following this reasoning, the existential comparability is valid. This is one explanation.

This explanation is a clear refutation for those who claim the earth weights as small as an atom in this galaxy, and that the heaven or the cosmos contain it. If this is the case, how could it be possible to maintain that the heavens and the earth are included in the heavens to begin with? Pay attention to this subtilty because it is crucial. Linguistically speaking, it is neither rhetorically accurate nor is it an Arab tradition. One note to bear in mind is that Allah couples opposites such as jinn and humans, east and west, sun and moon, day and night, as is the case with the heavens and the Earth. If the heaven is not the opposite of the Earth, why does Allah mention them together in multiple instances?

The second explanation addresses the verses including the heavens and the Earth and that is between them [is a place] for Allah's creation. The contemporary classification of the heavens as the space is incorrect, because the latter is empty and hosts nothing. Indeed, Allah does not create nothing or emptiness, nor did Allah create the heavens and the Earth out of amusement or in vain. Everything harbors a rationale behind its creation, and Allah knows best. This reasoning may be extended to refute atheists who understand evolution inaccurately. They say, "What is the need for a God when the Evolution theory explains everything?" This verse is a strong refutation of their claim.

The third explanation is that the phrase "what is in between them" refers to this formidable barrier Allah designated to separate the heavens and the Earth (or the ceiling sky, if you wish). The "in-between" space includes everything on the Earth, in the atmosphere of the sky, and anything that typically exists between the lowest heaven and the earth. In the language of the People of the Book, this is called the "Vault" or the "Firmament." This issue retains a profound significance for them as much as it does for us, because it is an essential component of existence. In addition, it is a proof that Allah created the heavens and the Earth after they were a single entity.

The fourth explanation is that the human, in general, lives in between them. Neither can he penetrate the lowest heaven nor can he pass beyond and dive beyond the regions of the Earth. So, he is eternally destined to live "in-between them."

The fifth explanation is that when Allah says, **"Allah created the heavens and the earth,"** the individual may inaccurately think Allah refers to the currently known inanimate objects or only the high heavens and the Earth, without any relation to the living creatures like the human being, land and maritime animals, birds, and planets, and so on. The individual may think Allah created the universe and abandoned the living creatures, according to some philosophers. Perhaps the creatures came through Evolution, as proposed by naturalists and biologists. However, the objective of Allah's saying **"created the heavens and the earth"** is that He is the One responsible for creating everything between the topside and underside of the cosmos or existence, including the living creatures.

Let me highlight a remark that does not necessarily offer the same rea-soning at every instance but is reasonable at most of its instances, according to my study. Whenever Allah, Exalted and Glorified, mentions **"His is the dominion of the heavens and earth"** [Al-Hadīd:2], without saying "and what is in between them," it may be deduced that He has a dominion over the entire cosmos. There are some exceptions in some verses for reasons relevant to those verses. For example, when Allah speaks about the Christ, He says, **"They have certainly disbelieved who say that Allah is Christ, the son of Mary. Say, 'Then who could prevent Allah at all if He had intended to destroy Christ, the son of Mary, or his mother or everyone on the earth?' And to Allah belongs the dominion of the heavens and the earth and whatever is between them. He creates what He wills, and Allah is over all things competent"** [Al-māidah:17]. Stating "whatever is in between them" emphasizes that Christ is among those who live between the heavens and the Earth. He belongs to Allah without having any dominion whatsoever.

Allah, Almighty, said, **"To Him belongs whoever is in the heavens and the earth. And those near Him are not prevented by arrogance from His worship, nor do they tire"** [Al-Anbiyā':19]. Glory be to Allah! In this verse, Allah states that everything in the heavens and the Earth belongs to Him. In short, He refers to the cosmos in our contemporary terminology, **"And those near Him."** This part of the verse subtly maintains that He and the creatures near Him whom He only knows (could be the angels knowns to Him) are above this existence called "the heavens and the Earth." Hence, this explains the prophetic hadith, **"The throne is the canopy of creatures."** It also agrees with many of the reports attributed to the Messenger (peace and blessings be upon him).

The same applies to His saying, **"to Him belongs whatever is in the heavens and the earth"** [Al-Baqarah:116]. Each verse has its own subtle remarks and treasures of wisdom that require an independent study. However, to stick to the objective of this book, the point of view I am proposing here about the heavens and the Earth corresponds perfectly and beautifully with all the Quranic verses, the sound Prophetic hadiths, and Torah. It is as if these pearls are put together in one thread to make up a beautiful necklace. Do reread what I previously stated and ready yourself for the what is coming. Perhaps you shall find the enlightenment every truth seeker desires. I hope you now recognize the difference between the "heavens and the Earth," "the seven heavens," "the heaven is a canopy / ceiling," "the atmosphere of the sky," and "the Earth."

4

EARTH AS A BED

ALLAH, EXALTED AND Glorified, said, "**[He] who made for you the earth a bed [spread out] and the sky a ceiling and sent down from the sky, rain and brought forth thereby fruits as provision for you. So do not attribute to Allah equals while you know [that there is nothing similar to Him]**" [Al-Baqarah:22].

This is the first reference to Earth in the Quran. The address is directed at people in general. Allah, Exalted and Glorified, explained that the Earth is a "Ferash" in Arabic which means bed, and the meaning of the bed is known to everyone. All the Arabic lexical meanings of the word "bed" lack the meaning or even the slightest implication of "spherical." On the other hand, a bed is usually flat to provide stability, sleep, and comfort laying down. Some exegetes cited this verse to argue Earth is stationary because it is a bed, which is clear. Someone may argue when a ball is gigantically enormous, it can be a bed just like a big mountain on which the individual can lay down and sleep, which is also correct. A large-sized object like a mountain can serve as a resting bed. My question is, is it linguistically plausible to call the mountain or the ball a bed? I do not think so.

In general, the verse lacks any implication to the sphericalness of the Earth. On the contrary, it implies flatness, which is why the reader is prone to think the Earth is flat. For this reason, some people who are influenced by philosophers maintain that this verse is not at odds with the sphericalness of the Earth.

This attempt of reconciliation is triggered by those arguing the Earth is flat, not spherical, because the explicit, sensible, and first thought to come is flatness.

In the commentary of *Mafātīḥ al-Ghayb*, al-Rāzī offers some interesting insights to the extent it attracted the imitation of many people afterward. He said, "The Earth as a bed is conditioned on the following: First: it is stationary otherwise it would follow a straight or a circular motion. If it had been moving in a straight path, it would not have never become a bed for us because anyone who leaps from a high position would not reach the Earth as it is falling into the abyss as well as the human. Only the Earth is heavier than the human, thus falling down faster than him and he cannot reach it. Accordingly, had Earth been an abyss, it would not be a bed and the human would not be above it. As for moving in a circular motion, we would not have benefitted from it. For example, if the Earth is moving east and the human wants to move west, the human would remain still and cannot reach his destination taken in consideration the Earth is faster in movement. But since the human can reach his destination, we can conclude that the Earth does not move in either a straight or circular motion; it is stationary, but the reasons are differed over among scholars. One view argues that the Earth is bottomless, thus void of a landing site and incapable of falling. However, this opinion is flawed because the calculability of objects has been established. Second: those who acknowledge the calculability of objects maintained that the Earth is not spherical, but it is seemingly half spherical whose cam is on top and its surface at the bottom. The surface lies above water and air and infused with them because it is heavy and spread out, at which case it floats on the water though it precipitates when condensed. This view is flawed for two reasons: 1) attempting to locate the reason for the stillness of water and air is identical to finding the reason for the stillness of Earth; Second: why is there a spread outside of the earth surfacing on the water while the other is came?

"Third: this opinion ascribes the motionless of the Earth to the gravitational pull of the water mass levelling it equally from all directions, thus forcing it into the middle. Still, this opinion is flawed for two reasons: first, the smaller object is pulled faster than the bigger one otherwise why isn't the dust particles pulled towards the water; second, a proximity of position makes it more susceptible to be pulled, for a particle thrown up high is more susceptible to be pulled though it shouldn't have returned.

"Fourth: the water mass is propelling the Earth from all directions similar to inserting some soil into the middle of a bottle and spinning it on its axes rapidly. The soil will stand in the middle of the bottle owing to the equal force propelled from all directions. This opinion is also flawed for five reasons. First: assuming this force is in such magnitude, why has not anyone felt it? Second: why does not this propelling force the clouds and wind into a specific direction? Third: why has it not made its movement to the west easier than to the east? Fourth: the motion of heavy objects slows down the bigger it becomes, because it is only natural that bigger objects are propelled slower than smaller ones. Fifth: the motion of heavy objects must be faster when sliding down at the starting point than when settling, because it is farther from the water mass at the start point.

"Fifth: The Earth is located in the middle of the water mass, as held by Aristotle and the majority of his followers. This opinion is weak because all bodies share equal corporeality. Hence it must be allowable to allocate some of them with a specific character for which that state is required, at which case it lacks a free doer.

"Sixth: Abu Hāshim argued that the lower half of the Earth has upward-directed foundation while the upper half has downward-directed foundations, and the mutual propelling of both foundations is forcing the Earth to be stationary. However, to assume each half has a special characteristic cannot be conceivable unless there is a free doer. Accordingly, the motionless of the Earth is decided by Allah, Almighty, alone. Look at the Earth! You will find no hanger on top of it nor a brace underneath it. As for the nonexistent hanger, it would require an endless number of hangers if we assume it has a hanger. Following this reasoning, it is also invalid to say there is a brace underneath it. The only remaining option is that there must be someone who holds it by His power and choice. Allah, Almighty, said, '**Indeed, Allah holds the heavens and the earth, lest they move away from their place. And if they should cease, no one could hold them [in place] after Him. Indeed, He is Forbearing and Forgiving**' [fāṭir:41].

"Second condition: in order for the Earth to be a bed for us, it must not be extremely solid like a stone, since sleeping or walking above it pains the body. In addition, if the Earth were to be from gold, for example, agriculture would not be possible, nor buildings would be erected because it would not be feasible to dig and lay foundation. Likewise, the Earth must not be extremely soft like the water at which the leg dives.

"Third condition: The Earth cannot be so transparent because a transparent object absorbs light. An object of such quality cannot receive heat from the planets and the sun, which would make it extremely cold. However, Allah made the Earth dusty to receive light and heat, and thus becoming habitable as a bed for animals.

"Fourth condition: the land must be distinct from water because land is dispositioned to be under the water, at which case the oceans would submerge the land, making it unfeasible for us to live. But Allah transformed the disposition of the land and made some of its masses above the water like the island so that it would be habitable as a bed for us. Some people claimed that the Earth must not be spherical if it were a bed, citing this verse to prove that Earth is not spherical. However, this assumption is very unlikely because a piece of a gigantic ball becomes like a spread-out surface providing stability. More proof can be found in the fact the mountains serve as pegs for the Earth providing stability; this opinion is more probable, and Allah knows best."

Much of what is stated by al-Rāzi contradicts modern scientific theories about the Earth. His elaboration and refutations are in favor of the flatness of the Earth more than its sphericalness. The objective of this long quote is to point out the glaring contradictions between al-Rāzi and modern theories. It is so apparent it requires no explanation, particularly to exhibit how the Earth is stationary. As for showing the downfall of imitation by some exegetes, al-Bayḍāwy imitates al-Rāzi in many parts on his commentary, as if he is copying his words.

Al-Bayḍāwy (may Allah have mercy upon him) refuted those who cited the verse as a proof that the Earth is flat, "This does not entail it is flat because its spherical shape, gigantic size, and broad mass are not at odds with being a resting bed." Furthermore, many commentators have quoted al-Rāzi without the appropriate reference or verification. Contemporaries of today should take into consideration that the era in which al-Rāzi lived was full of theological and philosophical debates, not to mention the scholarly background of al-Rāzi, to uncover how al-Rāzi went beyond the literal meaning of the Quranic verses.

Currently, we find people who associate Islam with terrorism, which drives pseudo-scholars to interpret verses such as, **"And fight against the disbelievers collectively as they fight against you collectively"** [Al-Tawbah:36], as encouraging violence and terrorism. If a non-Muslim perpetrated an act of terrorism, they say "terrorism has no religion." Many people associate Islam with terrorism owing to the media pressure. Like any scholar, al-Rāzi was subjected to various kinds of pressures faced by any regular individual, and there is no might nor power except by Allah.

In my opinion, the best imitator of al-Rāzi is Abū Ḥayyān al-Andalūsī, author of al-Baḥr al-Muḥīt, who stated, "Some of the astrologers cited Allah's saying, **'[He] who made for you the earth a bed [spread out],'** as a proof that the Earth is flat, not spherical, otherwise the seawater wouldn't have been stable. His usage of this verse as a proof for his position is invalid because the verse indicates the Earth is neither spherical nor flat. It only indicates that people use it to rest as they do on their beds, whether it is shaped flatly or spherically. Resting is made possible due to the wide distance between its diameters and its broad mass." He continued, "It is possible that a portion of spherical shape object to host water if the object is stationary and uncircular. In the case it is circular, it is habitually impossible for water to settle in such a spherical shape."

What interests me in the above observation of Abū Ḥayyān is his quote of some astrologers using this verse to prove the Earth is flat. He called them "astrologers," not just traditional exegetes. My second observation is that he maintained this verse cannot serve as a proof for either the flatness or the sphericalness of the Earth. He also believes the Earth is stationary. The reason some exegetes have not denied the idea of the sphericalness of the Earth, in spite of the suggested flatness indicated by the literal meaning of the verse, is based on the fact that the Throne engulfs the world. They differed on this issue, and a significant production of theological literature ensued. To reconcile both the ideas that the throne engulfs the world and at the same time is a ceiling for it, they opted to choose the sphericalness of the Earth at the center of the circle engulfed by the Throne.

At any rate, scholars have cited the verse in question as a proof for the flatness of the Earth, but after al-Rāzi, that difference of opinions increased. Al-Rāzi, in a manner of speaking, opened the way to increasing probabilities since Earth was believed to be flat by most of the scholars preceding al-Rāzi. Al-Nasafī, for example, is one of the scholars who came after al-Rāzi. In his commentary *Madārik al-Tanzīl wa Haqā'iq al-Ta'wīl*, he said, "The verse has no indication whether the Earth is flat or spherical, since the presence of a resting place is possible in both scenarios." He shares the same opinion with the above mentioned author of *al-Baḥr al-Muḥīt*.

In his commentary *Baḥr al-Ulūm*, al-Samraqandy commented on this verse, "Linguists said that Earth serves as a carpet for the world." Dear reader, have you ever heard of a spherical carpet? I myself have never heard of such thing in my entire life, not even in the fantasy adventures of *Sindbad*, *Aladdin*, or the *One Thousand and One Nights*. In the writings of some Sufis, we can draw a similarity between them and the statement of al-Samrqandy. For example, Ismāʿil Haqqy stated in his commentary *Rawḥ al-Bayān fī Tafsīr al-Quran*, "The Earth is the carpet of the world surrounded by the great ocean."

For reasons of imitation, we find some exegetes include a refutation of those claiming the Earth is spherical in one instance while quoting a refutation of those claiming the Earth is flat in another instance, may Allah aid the editors of the early scholars. Furthermore, in the commentary of the famous Shi'i exegetes al-Tabrasy, *Majmaʿ al-Bayān*, he commented on the verse, "It means a carpet above which you can settle, rest, and act as you wish. Such cannot be possible unless the Earth is spread out and eternally stationary."

In short, this verse has been used by a large number of scholars to prove the Earth is flat. Likewise, a significant number of prestigious scholars have not denied the possibility that Earth is spherical. However, none of them held that the Earth is spherical and mobile, since many of them have followed the perspective of some of the Greek philosophers. In any case, the perspective of al-Rāzi and others destroys the modern scientific theory. Therefore, I do not know why some contemporaries use the statements of exegetes as evidence that they preceded the scientific society in producing conclusions, resolutions, and necessities of the modern sciences. Is it due to slacking in research and verification? Or is it due to misunderstanding and narrow horizons? Talk about a rock and a hard place!

5

SKY IS A CEILING

I EARLIER ADDRESSED THE verse, "**[He] who made for you the earth a bed [spread out] and the sky a ceiling and sent down from the sky, rain and brought forth thereby fruits as provision for you. So do not attribute to Allah equals while you know [that there is nothing similar to Him]**" [Al-Baqarah:22].

In this chapter, I will address the issue of the ceiling sky extending like a dome above the Earth, unlike what is established by the scientific journals that the sky is a space, and any ceiling is made by adding a brick above the other. Let us explore some of the statements made by exegetes (may Allah have mercy upon them) at which they show the sky as a ceiling like a dome, starting by ibn 'Abbas who is nicknamed "the Interpreter of the Quran." He said, "Every heaven is stratum above the other like a dome while the sky of this world is attached to the Earth from its edges." One statement is attributed to other companions that says, "[Allah built the sky] above the Earth like a dome." Wahb ibn Munabbih said, "Some edges of the sky are enclosing the Earth while seas are similar to the edges of a pavilion." Iyās ibn Mu'āwiya said, "The sky is arched above the Earth like a dome." In his commentary, Abu al-Layth noted, "The sky is a dome encircling you, and every heaven forms a stratum above the other like a dome, while the sky of this world is attached from its edges to the Earth." Ibn Abi Hātim related that al-Qāsim narrated from Abu Burda, "The sky does not take a square shape but arched like a dome seen by people as green."

On this topic, al-Tabary reported many narrations on his commentary, some of which are relevant to this issue in hand. He (may Allah have mercy upon him) said, "Mūsaibn ibn Hārūn related to me saying ʿAmr ibn Hamād narrated to us from Asbāṭ from al-Suddy in a report he mentioned from Abu Mālik and Abu ṣālih from ibn ʿAbbas and from Murrah from ibn Masʿūd and from many other companions of the Prophet (peace and blessings be upon him) commented on Allah's saying, 'and the sky a ceiling,' by, 'It means a ceiling sky above the Earth like a dome.'"

Bishr ibn Muʿādh narrated to us from Yazīd from Saʿīd from Qatāda commenting on the preceding verse, "Allah made the sky a ceiling for us." The famous exegete al-Qurtubi (may Allah have mercy upon him) said, "The [status] of the sky is like the ceiling for the house." In his commentary, al-Kashshāf', al-Zamakhshary, the Muʿtazlite (may Allah have mercy upon him) said, "[Allah] created the Earth that provides them with stability and a habitat they necessarily need. It is like a courtyard and a bed. Then He created the sky, which is like an installed dome and a tent set up above this Earth." Similar words are echoed by the famous exegete ibn Kathīr, "The sky is a ceiling." Al-Bayḍāwy said in his commentary, "The sky is a ceiling arched like a dome above you."

Al-Shawkānī, one of the late exegetes, said, "[Allah] made the sky like an arched dome and a ceiling for the house they live at." In his commentary al-Dur al-Manthūr, the encyclopedic exegete al-Siyuṭi said, "[Allah] built the sky above the Earth like a dome, serving as a ceiling for the Earth." In Naẓm al-durar fī tanāsub al-āyāt wa-al-suwar, al-Biqāʿi said, "The sky is a ceiling like tent encircling the habitable area."

There are some nice preaching words regarding this verse though it is akin to reality. It is mentioned in the commentary of al-Khāzin, Lubāb al-taʾwīl fī maʿānī al-tanzīl, "If the reflecting individual were to ponder on the world, he would find it like an inhabited house containing everything one needs. The sky is raised like a ceiling for the house, the Earth is spread out like a mat, the stars like lanterns, and the human being is like the owner of the house."

One interesting conclusion can be drawn from the verse. Since Allah, Almighty, said, **"[He] who made for you the earth a bed [spread out] and the sky a ceiling,"** this indicates the Earth was created beforehand, as were the Heavens, except they were not a bed or spread out. This supports other verses in this Glorious Book such as, **"Then He directed Himself to the heaven while it was smoke"** [Fuṣṣilat:11]; and, **"And after that He spread the earth"** [Al-Nāziʿāt:30], Al-Tabary noticed, like others, "The ceiling is attributed to the sky rather than the Earth because the former was created after the latter." The issue of whether or not the heavens were created before the Earth will be discussed in a separate chapter.

Pondering on these verses makes one realize the sky is an actual ceiling, not merely a metaphor. Allah, Almighty, said, **"[He] who made for you the earth a bed [spread out] and the sky a ceiling;"** and, **"And We made the sky a protected ceiling, but they, from its signs, are turning away;"** and, **"Have they not looked at the heaven above them— how We structured it and adorned it and [how] it has no rifts?"** and, **"And when the heaven is split open and becomes rose-colored like oil"** [Al-Rahmān:37]; and, **"And the sky will split [open], for that Day it is infirm"** [Al-ḥāqah:16]; and, **"And [mention] the Day when the sky will split open with [emerging] clouds, and the angels will be sent down in successive descent"** [Al-Furʾqān:25]; and, **"And when the sky is opened"** [Al-Murʾsalāt:9]; and, **"Are you a more difficult creation or is the sky? Allah constructed it"** [Al-Nāziʿāt:27]; and, **"When the sky breaks apart"** [Al-Infiṭār:1]; and, **"When the sky has split [open]"** [Al-Inshiqāq:1].

All of these descriptions prove the existence of a physically built body, not a space. Most of the words describing the shape of the Earth in the Quran indicate it is flat and so are the words describing the sky indicate it is a ceiling, which is echoed as factual in the Torah. I will discuss the part about the Torah in its separate chapter.

This ceiling sky has gates. Consider the saying of Allah, Exalted and Glorified, **"And [even] if We opened to them a gate from the heaven and they continued therein to ascend"** [Al-Hijr:14]; and **"And the sky is opened and will become gateways"** [Al-Nabaʾ:19]; and, "Then We opened the gates of the sky with rain pouring down" [Al-Qamr:11]; and, **"Indeed, those who deny Our verses and are arrogant toward them—the**

gates of Heaven will not be opened for them, nor will they enter Paradise until a camel enters into the eye of a needle. And thus do We recompense the criminals" [al-Aʿrāf:40]. The gates are repeatedly mentioned in the Prophetic hadiths. The fact that the sky has gates and windows is very common in Jewish and Christian traditions as well. In many museums, if you wish to visit, there are many paintings depicting people's desire to enter through the gates of the sky guarded by angels. The point I seek to emphasize by referencing the gates of the sky is that the sky is a ceiling, not an empty space, thus featuring the logical understanding that every ceiling has gates. Actually the word ceiling may not be the appropriate translation of the Arabic word "Bena'a" which literally means construction, I am using the words ceiling to be aligned with common English translation of the Quran which instead of using construction used ceiling which may actually gives a different meaning.

6

EARTH IS SPREAD OUT

A LLAH, ALMIGHTY, SAID, **"And it is He who spread the earth and placed therein firmly set mountains and rivers; and from all of the fruits He made therein two mates; He causes the night to cover the day. Indeed in that are signs for a people who give thought"** [Al-Ra'd:3].

The literal meaning of this verse indicates that Earth is flat, given we have never heard of a spread-out ball. Have you ever heard of a spread-out ball kicked by boys or footballers? But after the increase of the believers in the sphericalness of the Earth theory, the widespread popularity of the idea, the belief in it by all people, young and old, and its standardization in the school curricula as the undeniable truth, some exegetes attempted to reconcile between the Quranic verses and the statements of astronomers, which turned out to be decent, just as al-Rāzi did before. Their reconciliation, however, goes beyond the literal meaning of the verses. Would it have been burdensome had they harmonized the literal meaning of the Quran with the cosmic fact informing of the flatness of the Earth? Why hadn't they made one possibility of the flatness of the Earth? Unfortunately, the psychological pressures they faced from all directions, like today, made burdensome upon them to say, "Your conclusions are wrong." The constant anxiety clouds the individual's judgment. At any rate, let us explore some of the statements by some exegetes.

First, there are some reports attributed to the companion ibn 'Abbas, nick-named "the Interpreter of the Quran," in which he said, "[Allah] spread it out on the water." In other unverified reports related by Abu Nu'aym, Wahb ibn Munabbih said, "Building civilization in this life in a desolated area is similar to having a pavilion in a sea." Ibn Abi Hātim related that ibn 'Abbas (may Allah be pleased with them) said, "Earth is [divided] into seven parts, six of which are inhabited by Gog and Magog while the last part is inhabited by the rest of the crea-tures." In the commentary of Muqātil ibn Sulaymān, he said, "[Allah] spread the Earth out from underneath the Ka'ba and extended the Earth beyond the Ka'ba for as long as the [journey] of two thousand years. He made its length equal to a journey lasting five hundred years; and its breadth equal to a journey of five hun-dred years." Al-'Aṣṣam said, "Spreading out something is to extend it endlessly."

Ibn Jarīr al-Tabary said, "Allah spread out the Earth and extended it length and breadth." Furthermore, al-Qurtuby said, "Allah spread out the Earth in length and breadth." He continued, "This verse includes a refutation of those claiming that the Earth is spherical, and its gates are collapsing over it. Ibn al-Rrāwindy claimed that there is a body under the earth escalating like the blasting wind coming in contact with the gliding down earth. Both bodies share the same mass and power. Others have maintained that Earth is composed of two bodies, one of which is a slope while the other is uprising, thus balancing each other out and setting the Earth at a stationary, balanced stature. However, Muslims and the People of the Book believe that the Earth is stationary and spread-out; its occasional motion is usually due to an earthquake."

Al-Fayrūzābādī noted on his commentary, "Allah spread out Earth on the water." Al-Tabarāni said, "He spread it out in length and breadth." Ibn Kathir (may Allah have mercy upon him) said, "Allah made it extended and spread out in length and breadth." Al-Farrā' said, "Allah spread it in length and breadth." In his 'Baḥr al-'Ulūm' al-Samraqandy said, "Allah spread the Earth out from underneath the Ka'ba above the water. It was wobbling with its inhabitants just as a ship is for its passengers, but Allah firmed it by mountains." Al-Māwarddy said, "In refutation of who claims that Earth is spherical like a ball, Allah spread it out to provide stability above it."

Ibn ʿAṭyya said, "The verse requires the Earth to be spread-out, not a ball; and this is the literal indication of the Shariah texts." His works speak truly of the fact of the matter despite the attempt of al-Alūsi, the famous exegete, to rebuke this statement. Definitely, al-Alūsi would not have done so unless he was already imitating al-Rāzi in many of points he proposed, including this one in question. Al-Alūsi was certainly overwhelmed by the findings broadcasted widely about the shape of the Earth in his days. As for al-Rāzi who supported the sphericalness of the Earth, he listed the different views before proposing his. He said, "Some people claimed that the Earth was firstly spherical but Allah spread it out from underneath Mecca toward such-and-such. Others said it was one piece by Jerusalem but Allah commanded it to go toward such-and-such." The objective here is to show there was a significant number, if not seemingly the majority, maintaining the Earth was flat, until al-Rāzi proposed his idea that was influenced by the widespread idea of the sphericalness of the Earth during his era. He could not push back against it, owing to the persistent pressure imposed by the theory and its proponents. Hence, he went beyond the literal meaning of the Quranic verses, just like what imitators of the West do. They accept any piece of information relayed by the scientific society and change the literal meaning of the Quran to correspond with the dictations of the scientific societies.

Ibn Juzayy l-Gharnati said, "The [Quranic verse] entails that the Earth is spread-out, not spherical. It is the literal meaning conveyed by the [texts] of Shariah. It is possible that the meaning of spread-out branched off from sphericalness, because every piece in Earth is extended on its own while the sphericalness is for the Earth as one piece."

In his commentary, Abu al-Layth said, "Allah spread the Earth out from underneath the Kaʿba above the water. It was wobbling with its inhabitants just as a ship is for its passengers, but Allah firmed it by mountains." In some of the Sufi reports, before the creation of the heavens and the Earth, Allah, Almighty, sent out stiff wind in the water. They clashed with each other, which left a protruded rock at the location of the House (i.e., Ka'ba) in the shape of a dome. At this location, Allah, Almighty, broadened the Earth from its length and breadth. The origin of the Earth is at the Ka'ba, at the heart of the inhabited land. The center of the Earth, at both of its inhabited and desolate parts, is the dome of the Earth where time is equally divided between coldness and hotness. Day and night are timely equally. The origin of the clay of the Messenger of Allah (peace and blessings be upon him) is taken from the eye of the Earth at Mecca, and after there were waves with the water, it throws that piece of clay to his death's location at Medina, which is why the Prophet (peace and blessings be upon him) was buried there. There is a saying that goes, "The land is our bed and our mother. It is where we live and where we die."

In *Ta'wīlat Ahl al-Sunnah*, Al-Matūridy commented on **"And it is He who spread the earth,"** as follows, "It means He extended it and placed firm amounts therein. It was reported to have been extended above the water. It was wobbling with its inhabitants just like a ship but Allah made it firm by heavy mountains that provided it with stability and firmness. It was also mentioned that the Earth was extended on the air but was firmed, too, by mountains. However, assuming this was the case, the Earth's firmness and stability shouldn't have been provided by the mountains, because the Earth and mountains are naturally low and steep in both air and water; the more mountains, the more steepness and lowness, thereby ruling out the very idea of stability and firmness. On the other hand, stability and firmness can be provided by something naturally high that resists any steepness and lowness, unless it is argued that mountains were not low nor steep but instead wobbly, following Allah's, Almighty, saying, 'And We placed within the earth firmly set mountains' [Al-Anbiyā':31].

"According to this reasoning, mountains did provide stability and firmness for the Earth and countered the wobbly and slanty nature of the Earth. It is possible this was mentioned to highlight Allah's kindness and power, since He, Almighty, firmed the Earth with something sharing its steepness and lowness, and Allah knows best. Allah's saying, '**spread the Earth**' means He created it spread-out. It was not a one piece then He spread it out like the elevation of the sky."

In his *Taysīr al-Tafsīr*, al-Iṭfīsh al-'Ibādi said, "This verse proves that the Earth is flat; so is Allah's saying, '**And after that He spread the earth**' [Al-Nāzi'āt:30]. There is indication of sphericalness because its spread-out shape is due to its gigantic size so much that each piece's shape shows its surface.

In relation, the evidence of philosophers on this regard are alien. As for the literal meaning of 'spread the earth,' it indicates that Earth existed without the spread-out shape and then it was spread out, which is unobjectionable. If interpreted that it was created spread-out, it would mean that the spread-out shape is its one and only shape since Allah, Exalted and Glorified, created it. It may also mean that Allah created it extended like the narrowness of the mouth of the well."

To conclude this exploration, I would like to quote a fine observation by al-Khāzin in his commentary, *Lubāb al-ta'wīl fī ma'ānī al-tanzīl*. He said, "Allah spread it above the face of the water. It is said that the Earth was one piece but Allah spread it out from underneath the Sacred House. This view is only plausible if it is assumed that the Earth was flat like palms of the hand. According to astronomers, the Earth is spherical. It can be reconciled that if the Earth's shape is spherical, each piece of it will be seen as extended, like a huge surface, given the gigantic size of the Earth. However, Allah, Almighty, said that He spread it out and extended it, all of which leads to the conclusion that the Earth is flat. Allah, Almighty, is most Truthful and provides more powerful evidence than astronomers."

It is obvious now that most of the exegetes understood the literal meaning of the verse to indicate that Earth is flat, not a spherical one floating in the wide space. Also, Allah said about extending the Earth, **"And the earth—We have spread it and cast therein firmly set mountains and caused to grow therein [something] of every well-balanced thing"** [Al-Hijr:19]. The statements of exegetes regarding the verse in chapter al-Ra'd have been mentioned earlier. Likewise, their statements regarding this verse in chapter al-Hijr will be explored as follows.

Qatādah said, "It was extended from Mecca because it is the mother of cities." Muqatil said, "The verse means we have spread it out in a distance [equaling] the journey of five hundred years at its length, breadth, and width. Allah spread it out from underneath the Ka'ba." Ibn Jariīr said, "Allah has extended the Earth as in spreading it out." In addition, al-Qurtuby said, "This is also one of Allah's favors and one of the indications of His complete power." Ibn 'Abbas said, "We (i.e. Allah) extended it above the face of the water, according to

Allah's saying, '**And after that He spread the earth;**' and, '**And the earth We have spread out, and excellent is the preparer**' [Al-Dhāriyāt:48]. This was a refutation of those who claim that the Earth is a spherical shape."

In his *Madārik al-Tanzīl wa Haqā'iq al-Ta'wīl*, al-Nasafī said, "We (i.e. Allah) extended it from underneath the Ka'ba. The majority of scholars agree that Allah, Almighty, spread it out above the face of the water." Al-Shawkānī said in his commentary *Fath al-Qadīr*, "We (i.e. Allah) extended and spread out the Earth, as in His saying, '**And after that He spread the earth;**' and, '**And the earth We have spread out, and excellent is the preparer.**' This is a refutation of those who claim that the Earth is a spherical shape."

As usual, al-Rāzi opted to support the sphericalness of the Earth. He said, "Does '**And the earth We have spread out**' indicate that Earth is flat? Yes, it is, because Earth is a gigantic ball assuming it is one. Upon looking at each small piece of it, it looks like a flat surface. Hence, this problem is solved." He acknowledged that the literal meaning of the verse demonstrates it is flat, but he then presumes that the Earth is a gigantic ball, only to reconcile between the Quranic verses and the widespread idea of the sphericalness of the Earth in his era and the succeeding eras. Al-Sha'raawi noted, "This is a gesture that Allah, Almighty, is guiding us to see. If the Earth were to take the shape of a square, triangle, or a rectangle, it would have an edge and an end. However, we find Earth extended when we walk above it, which is why it must be spherical."

It is easy to refute al-Sha'raawi by saying, "Who knows the borders of the Earth?" It is linguistically acceptable to call the Earth flat if it is massively extended, just like the word "eternity." As a result, al-Sha'raawi's statement is debatable. The fact that Earth has an edge or an end is not contradictory to being flat whose end cannot be reached by the individual. In this book, I have designated a chapter refuting, from the Quran, the evidence proposed by the proponents of the sphericalness of the Earth. I will elaborate on each verse to expose the shortcomings of the evidence of that viewpoint.

Al-Khāzin said, "We (i.e. Allah) have extended it above the face of the water. It is said that it was extended from the starting point underneath the Ka'ba; this is popular opinion of exegetes. As for astronomers, they claimed that the Earth is a gigantic ball, one part of which is submerged under water while the other part is above the water. They reasoned Allah's saying, '**And the earth We have spread out**' that when a ball is gigantic, each piece of it will look like a huge flat surface. Accordingly, this proves that the Earth is flat and it is a ball. Exegetes have discounted this reasoning on the basis of the very statement of Allah informing us it is flat in His Book. Had it been a ball, Allah would have told us since He is the One who knows the true meaning of His words and how He extended the Earth."

7

SEVEN HEAVENS

A**LLAH, EXALTED AND** Glorified, said, "**It is He who created for you all of that which is on the Earth. Then He directed Himself to the heaven, [His being above all creation], and made them seven heavens, and He is Knowing of all things**" [Al-Baqara:29].

What are the seven heavens? How does modern science today interpret the seven heavens mentioned in the Quran as well as in many sacred Scriptures and texts? The advocates of modern science offer no cogent interpretation that is congruent with the Quran. More importantly, where is the sky now? All given answers merely define the heaven as the blank, empty space. Alas, some contemporary Muslim scholars subscribed to this notion, contending that space per se is the heaven. The crux of the arguments of most of them is that whatever is high above you is deemed the sky. In this regard, I would like to point out that Allah has provided us with the qualifications of the sky. As I have stated earlier in the chapter on the ceiling sky, Allah described the sky for us in His Scripture by saying, "**and the sky a ceiling**" [Al-Baqara:22]. Therefore, the heaven is a ceiling. However, I did not find in the Arabic usage that "space" means "ceiling." If it were a blank space, how did He structure the heaven into seven heavens? This is outrightly incongruent with the conclusions of modern science. In a similar vein, He said, "**By the heaven containing pathways**" [Al-Dhariyāt:7]. He also said, "**And constructed above you seven strong [heavens]**" [Al-Naba':12]. He further asserted, "**And the heaven We constructed** with strength, and indeed, We are [its] expander" [Al-Dhariyāt:47]. He also said, "And We made the sky a protected ceiling, but they, from its signs, are turning away" [Al-Anbiyā':32]. These are but a few Quranic verses that clearly show that the sky is a ceiling and that it is massive, firm, and solidly built.

Some contemporary orators relentlessly try to reconcile the conclusions of academic periodicals and scholarly journals with the verses of the Quran. In so doing, they come up with ideas that are heavily influenced by what Westerners reiterate. For instance, they argue that this cosmos of ours is no more than the space in between the sky and earth, and that the sky stands above all of this cosmos including its galaxies, stars and so on. Well! Then how about the stars that we see? Did not Allah describe what he had created in the sky in His Scripture and how He adorned and guarded it, and so on? He even stated clearly what is in between the sky and earth. Allah, Exalted and Glorified, said, "and the clouds controlled between the heaven and the Earth" [Al-Baqara:164]. On commenting on this verse, al-Fayrūzabādī stated in his commentary that, "the seven heavens built flat on earth." Ibn 'Atiyya in his commentary, Al-Muḥarrar al-Wajīz, opined that "the term 'sawwahunna' is said to mean 'He made them equal.' It is also argued that it means 'He made their surfaces flat.'" These above mentioned commentaries encompass untold implicit conclusions that I shall elaborate on the following pages of this book.

Here, I would like to give you a rundown of what some commentators maintained regarding the seven heavens.

Al-Fakhr al-Rāzī stated, "Know that the Quran here asserted that there are seven heavens. Astronomers contended that the closest celestial body to us is the moon, then Mercury above it, then Venus, then the sun, then Mars, followed by Jupiter, and followed by Saturn." He then added, "Know that astronomers claim that the planets are nine. The first seven of them are the one we mentioned earlier. The eighth planet is the orbit in which these planets revolve. As for the ninth celestial body, it is the most superior planet which revolves approximately one full spherical every day and night."

Al-Baydāwī, however, commented that, "if it is argued that astronomers had proven that there are nine celestial planets, I would argue back that their contention is still dubious and open to question. Even if their contention were to be arguably true, the Quranic verse does not overrule the existence of more celestial bodies. Moreover, if the throne and the seat were to be added to the seven heavens, then there will be no contradiction." Just think about how al-Bayḍāwī (May Allah have mercy on him) tried to confute the question that astronomers have proven there are nine planets. He seems to be pointing out a confusion or contradiction between the Quran and what astronomers have come up with. In other words, it seems that the scholars of Quran exegesis and others in the discipline took for granted that the heavens are the planets. Furthermore, think about how he sought to reconcile the Islamic sciences with the conclusions of astronomers, and, therefore, he added the throne and the seat to the planets.

In his commentary, *Gharā'ib al-Quran wa-raghā'ib al-furqān*, al-Qommī al-Naisaburī mentioned that the heavens are the celestial bodies that stand high above us "in their entirety and nobody from the early generations or future ones knows for sure the number of the heavens as they are by means of reasoning or Scriptural prooftexts, **'and none knows the soldiers of your Lord except Him. And mention of the Fire is not but a reminder to humanity'** [Al-Muddathir:31]." I do not think al-Naisaburī understood anything about celestial bodies beyond the cosmos as the modern scientists do today. In this regard, I would like to note that some exegetes were of the opinion that the seven heavens are finite; namely, they are exactly seven heavens and nothing more. In actuality, their view is not far from true and can easily be corroborated.

In his commentary *Maḥāsin al-Ta'wīl*, Muḥammad Jamāl al-Dīn al-Qāsimī remarked that, "because of that, an ostensibly clear miracle looms large because, in the era of Arab civilization and progress when Muslim scholars promulgated knowledge to every nook and cranny of the world, astronomers only knew five celestial planets in their same Arabic names we know today. These planets were Mercury, Venus, Mars, Jupiter and Saturn. They understood them as the heavens mentioned in the Quran. However, when it was not possible to reconcile the five planets with the seven heavens in terms of number, they resorted to adding the sun and the moon to make the planets seven even though the Quran clearly declares that the seven heavens are different from the sun and the moon. Allah, Almighty, said, '**It is Allah who erected the heavens without pillars that you [can] see; then He established Himself above the Throne and made subject the sun and the moon, each running [its course] for a specified term**' [Al-Ra'd:2]."

The term *sakhara* (made subject) is an indication that sets the sun and the moon apart from the seven heavens. Therefore, the commentators, who are not familiar with astronomy, held the view that the sun should not be counted as a heaven, and the same applies to the moon, too. This is because they know that the seven heavens are inhibited but the sun is a burning fire. Therefore, their commentary regarding the meaning of the heavens rested on these suppositions. When an unknown celestial planet was later discovered with the help of the telescope, they called that planet Uranus and then another one was discovered and it was dubbed Neptune, and that made the number of planets seven. These discoveries, which took place 1,200 years after the demise of Prophet Muhammad (peace and blessings be upon him), underscore the miraculous nature of the Quran and substantiate the prophethood of the one to whom the Quran was sent down, Prophet Muhammad (peace and blessings be upon him). In terms of in-depth scrutiny, there are a number of valid objections to this view.

The renowned Ibadi sheikh al-Khalīlī, who is a contemporary exegete, took issue with the aforementioned view in his commentary, *Jawāhir al-Tafsīr*. He maintained, "Allah stated in this verse as well as in numerous other verses that the heavens are seven. We cannot help but take what the All-Knowing, the All-Aware for granted. However, human beings are created curious and eager to know everything that is unknown to them. Therefore, exegetes unleashed their thinking in a bid to arrive at the intended meaning of the seven heavens. Not only did they diligently indulge themselves in that endeavor but many of them consulted the falsified accounts of the People of the Book as well. As a result, their Quranic commentaries came fraught with interpretations that do not conform to Allah's words and thus should be discredited, especially since there are clear indications and decisive proofs that make the falsity of the accounts they drew from very obvious.

Moreover, the commentators, past and present, who are acquainted with astronomy, interpreted the seven heavens in light of their limited knowledge of the science of astronomy. Unfortunately, they failed to provide interpretations that hold to Allah's intent for the elaboration of His signs to human beings regarding the mightiest of His creations. As a consequence, we find that earlier exegetes opined that the first heaven is the moon, the second is Mercury, the third is Venus, the fourth is the sun, the fifth is Mars, the sixth is Jupiter, and the seventh is Saturn.

Hence, we conclude that they understand the heavens to mean the solar system. Since some of the planets in this solar system were not known to them at the time and since the planets that they were known of them back then were less than seven, they added both the moon—albeit being ancillary to the Earth—and even the sun to make them seven. Al-Fakhr al-Rāzī espoused this interpretation and proffered a lengthy discussion to substantiate it in light of what was known to astronomers in his era. Interestingly, later commentators espoused the interpretations of their predecessors where they interpret the heavens as the solar system. However, they only differ with them in not counting the sun and the moon among the seven heavens. They do so because the moon is associated with the Earth and is not independent from it whereas the sun is the center around which all planets revolve. Instead, they added the two planets recently discovered, Uranus and Neptune. Ironically, they even considered their discovery twelve centuries after the revelation of the Quran as a proof of the miraculous nature of the Quran, given that it numerated the heavens as seven. Some of them also observed the description of the lowest sky in the Quran as one adorned with star-like lamps, as it is the case in Allah's saying, **"And We have certainly beautified the nearest sky with stars"** [Al-Mulk:5], and in His other saying, **"Indeed, We have adorned the nearest sky with an adornment of stars"** [Al-Ṣāffāt:6]. Consequently, such a description was understood to mean the celestial bodies revolving around the closest planet to Earth (i.e., Mercury). Put differently, the heavens are these planets as well as other celestial bodies that revolve around them in space. Notably, scholars past and present agree on excluding planet earth and did not count it among the seven heavens because the addressees live on it and from there, they observe other planets. Moreover, some of them argued that the seven heavens mean the seven orbits. However, this view was rejected, given that these orbits are mere virtual curved paths which these planets take as they revolve around the sun. As such, they are not real orbits. For me, all these views are flawed and incorrect. I am more inclined to understand the heavens in terms of the linguistic usage of the term and in light of the modern scientific discoveries."

As we have seen, some eminent scholars did not break away from the tentacles of imitation. While the argument of the former scholar lacks in-depth scrutiny, the latter interlocutor offered an unsubstantiated refutation and thus offered a faulty response that cannot be counted as a sound rebuttal. Indeed, Allah is the main Helper, and we seek refuge in Him from imitating without exerting the efforts required for scholarly scrutiny.

Muḥammad Rashīd Riḍā maintained, "Allah said, '**Then He directed Himself to the heaven**'" [Al-Baqara:29]. When it is said '*istawā ilā al-shai'ī*,' it means he turned specifically toward it, and nothing could turn him away from it. Al-Rāghib noted, "if the verb *istawa* came intransitive followed by the preposition *ilā* (to), it means 'getting to that thing either by oneself or by planning to do so.' The meaning is that His will turned towards the heaven as He stated in the chapter of Fuṣṣilat: '**Then He directed Himself to the heaven**'" [Fuṣṣilat:11], Allah's saying, "**and made them seven heavens**" [Al-Baqara:29], means He completed their creation of that smoke substance and made them seven fully completed, perfectly arranged heavens. This succession conforms to the one that was known for Jews from Prophet Moses (peace be upon him) that Allah, Almighty, created the Earth first and then created the heavens and the light. It is also possible to only hold to the literal meaning of the verse because creation is different from turning toward something. Do you not know that the human being in the phases of the sperm-drop and the clinging clot is considered a creation but not a full human being in the best form as it is the case when he is developed into another creation?

We shall elaborate more on this when commenting on the following verse, "**Have those who disbelieved not considered that the heavens and the Earth were a joined entity, and We separated them**" [Al-Anbiyā':30], and expound on the fact that the cosmos was one mass but then Allah split it into parts through creation and ordained its precise determination. It is possible then that the Earth and everything on it was created before heavens were structured into seven heavens. In fact, this is one of the secrets of creation that we know nothing about. Some might erroneously think this verse contradicts Allah's other saying on the creation of the heaven and its lights, **"And after that He spread the Earth."** [Al-Nazi'āt:30]. This could be responded to in two ways. First, the sequence here is not a temporal order but a sequence in terms of mention. This is well known in the lingual usage of Arabs and others. It is correct to say, "I did such-and-such for so and so, did him such-and-such favor, and then helped him out in such-and-such matter." It is the same as saying, "In addition to that, I helped him out," referring to doing him another favor concurrently and without intending the consequence of the two acts. Second, the creation of the heavens was followed by spreading out the Earth, that is, making it flat, leveled, and inhabitable. The mere creation of it and ordaining the substances of its inhabitants is not what is meant here, for Allah's creation and His ordinances on Earth have been ongoing and will continue for good so long as the Earth exists. The same applies to Allah's other creations.

In addition to that, I would like to add here that the term *Dahw* in the Arabic usage refers to trundling circular, rollable things such as walnuts, balls, and stones and throwing them. They describe rain as *dāḥī* (rolling) because it moves pebbles. The same applies to the one who plays with walnuts. In the Prophetic tradition narrated by Abū Rāfi, it reads, "I was playing with al-Ḥassan and al-Ḥussein with *al-Madāḥī*," which refers to spherical stones like discs into which they would dig a hole and throw these stones. If a person managed to get his stone to fall in the hole, he would win, and if not, he would lose. It was also mentioned in Lisān al-'Arab lexicon, and he said after that the term *dāḥī* refers to throwing a stone, a walnut, or the like. I add that what has been mentioned about the game of throwing a pebble in a hole is still known to young children in our countries and call it "*al-'Ukra* (Ball-throwing Game)" and some pronounce the Arabic name inaccurately changing its "*u*" sound to a "du" sound.

In his *Mufradāt al-Quran,* al-Rhāghib stated, "Allah, Almighty, said, '**And after that He spread the Earth**' [Al-Nazi'āt:30]. It means He moved it from its location as in His other saying, '**On the Day the Earth and the**

mountains will convulse'" [Al-Muzzammil:14]. This is similar to its usage in this example, "The rain *daḥa* (moved) the pebbles and the like." However, there is a difference between rolling the Earth and moving it from its position during creation and shaking it fiercely to destroy it shortly before the Day of Judgment. The intended meaning could be that *daḥaha* means He separated it as well as the heavens from the smoke substance that was one mass, thus an indication that the Earth is at least spherical or shaped like a ball in terms of sphericalness. And Allah knows best.

It is not far-fetched to argue that moving the Earth and rolling it mean that He, Almighty, made it revolve in its orbit thanks to His Omnipotence, **"each, in an orbit, is swimming"** [Yāsīn:40]. This does not contradict what has been stated earlier that it refers to spreading out the Earth, stretching it wide, and making it flat so that human beings and other creatures could live on it. Those who argue that its sphericalness and its flat surface remain at odds since they are two camps that espouse one view and take issue with the other's position. Those people cannot apprehend the ample flexibility of the Arabic language and of the religion because of their scant knowledge of both.

The crux of the matter is that Allah, Almighty, has created this earth and these heavens that stand high above us, and He did not make us witness their creation. He only informed us about them in order to provide indications to His Omnipotence, Wisdom, and the blessings He bestowed upon us. He did not do so to detail the account of their creation in an orderly fashion because that is not the objective of the Islamic faith. The beginning of creation is unknown and so is the consequence of creations. Yet, the structuring of the heavens into seven appears to have taken place after the creation of the Earth. It seems that the heaven existed, but they were not structured in seven heavens. That is why He spoke of turning toward the heaven and said, **"and made them seven heavens"** [Al-Baqara:29]. We believe that he has done that for reasons only He knows. He manifested that to us to ponder over, and whoever wants to know more should do so through exploring the cosmos. They should also look into the contributions of earlier seekers of knowledge about the cosmos and those who made breakthrough discoveries in its various matters. They should only accept conclusions corroborated with sound proofs and reject the flawed ones of the pseudo-scholars who come up with unfounded assumptions and suppositions. Suffice it to say that the Quran guided them to do so and made it permissible.

Al-Ṭāhir ibn ʿĀshūr maintained, "in this and other verses, Allah, Almighty, enumerated the heavens as seven and He knows it best as well as the intended meaning of it. However, the explicit meaning indicated by scientific guidelines is that the heavens are the great, higher celestial bodies, namely the planets revolving with the Earth in curved orbits in the solar system. This is substantiated by a number of facts. First, the heavens are mentioned together with the Earth in most occurrences in the Quranic verses. He even mentioned their creation in the context of creating the Earth and this ostensibly signifies that they are planets like planet earth. This is true for the orbiting planets. Second, the heavens were mentioned along with the Earth as manifestations of Allah's perfect gargantuan creations. Therefore, it is befitting to interpret them as these celestial bodies observed by the people and known for all nations, particularly because their elliptical revolution in orbits and gleaming radiance evidently bespeak the magnificence their Creator.

"Third, they were depicted as seven heavens, and astronomers have known the seven planets since the Chaldean epoch. Since then until the revelation of the Quran, subsequent generations of astronomers have been unanimous that they are seven. Fourth, these celestial bodies are the planets whose revolution in orbits lines with the rotation of the sun and the Earth. That is why contemporary astronomers dub them as the solar system. As a result, it is befitting to mention their creation together with the creation of earth. Moreover, some go as far as arguing that the heavens are the orbits but this interpretation is fallacious because the orbits are the curved rotation paths that revolving planets take in space. These are virtual lines that have no tangible existence in reality. Besides, Allah, Almighty, stated that the heavens are seven in this verse and elsewhere in the Quran. He also declared that the throne and the seat encompassing the heavens and He made all the heavens opposite to the Earth. This corroborates the argument astronomers proffered that the celestial planets are nine and these are their names in order in terms of their distance away from earth: Neptune, Pluto, Uranus, Saturn, Jupiter, Mars, the sun, Venus, Mercury and Vulcan.

"Technically speaking, astronomers deem the Earth as a celestial planet. In the language of the Quran, however, it is not counted as one of these planets because it is the place from which we observe other planets. Yet, the moon was counted in lieu thereof on the grounds that an ancillary to Earth. Therefore, it was enumerated as one of these celestial planets to make the structure of the heavens more intelligible. The fast and fixed rules for astronomers conclude that they are celestial bodies orbiting in the vast space around the Earth but such contention is dubious and without foundations. Perhaps Allah, Almighty, did not create the Earth in the form of the seven celestial planets and that is why He did not count it as one of them. Perhaps Allah, Almighty, set forth the heavens that are associated with the planet earth of ours."

The renowned Sufi Ibn 'Arabī noted in his commentary, "He proportioned them into seven heavens as per what lay people could observe, for the eighth and ninth are obviously the throne and the seat."

Hence, we realize that Quran commentators differed with regard to the interpretation of the seven heavens. If one were to ask me about my opinion regarding this issue or the conclusions I have arrived to, I would offer the following response: I have sundry ideas, but I have not found my way to the truth yet. Still, it seems to me that the seven heavens are layers one after the other or surfaces whose milestones or signs are known today as the seven planets widely known to human beings since time immemorial. These seven planets revolve in orbits, tracks, and levels that serve as signs of these heavens. To make this clearer, try to draw a rainbow with its seven colors on a globe, then label each color with the name of a planet of them to identify it and to make it clear that this heaven is of a different color from the one above it and so on. And Allah knows best.

The proof that the people at the time of the revelation of the Quran were well-acquainted with what the seven heavens are is that when Noah (upon whom be peace) reminded his community of Allah's blessings to them, he reminded them that Allah had created seven heavens. These blessings are cited in the chapter of Noah, "**What is [the matter] with you that you do not attribute to Allah [due] grandeur. While He has created you in stages? Do you not consider how Allah has created seven heavens in layers** and made the moon therein a [reflected] light and made the sun a burning lamp? And Allah has caused you to grow from the Earth a [progressive] growth. Then He will return you into it and extract you [another] extraction. And Allah has made for you the Earth an expanse that you may follow therein roads of passage" [Nūḥ:13–20]. He also stated in the chapter of al-Mu'minūn (the believers), "Say, 'Who is Lord of the seven heavens and Lord of the Great Throne?' They will say, '[They belong] to Allah.' Say, 'Then will you not fear Him?'" [Al-Mu'minūn:86–87]. It means they already knew about that, and this is the most cogent conclusion I have arrived at, thus far, "and say, 'My Lord, increase me in knowledge'" [Ṭaha:114]. It seems to me that Allah, Exalted and Glorified, reiterated His creation of the seven heavens in order to benefit human beings. This interpretation of this verse is novel, and you will find proofs in numerous other Quranic verses that corroborate my argument, as I shall explain later.

What is the point of what I have mentioned earlier? What exactly do I seek to highlight?

I want to pinpoint the fact that human beings on earth were familiar with the seven heavens, but the humans today barely know what the nearest heaven is! This is because they espouse the conclusions of the scientific community and thus fell in sheer confusion.

8

SUCCESSION OF THE CREATION

OF THE HEAVENS AND EARTH

L ET US GET back to the question of the creation of the heavens and earth
and which one was created first. Allah, Exalted and Glorified, said, "**It
is He who created for you all of that which is on the Earth. Then
He directed Himself to the heaven, [His being above all creation],
and made them seven heavens, and He is Knowing of all things**"
[Al-Baqara:29]. My observation is that Allah created the seven heavens after
creating the Earth. Nonetheless, those who espouse the conclusions of modern
science will take issue with this observation because, according to them, the
Earth came into being as a result of the Big Bang. After that, the heaven and
space were formed. If they were to argue that the Big Bang theory relates to the
cosmos in between the heavens and earth only, we will ask them, "What was
beyond this cosmos? Was there non-existence, and then Allah created seven
heavens? Yet, proffering a cogent answer to this question is close to impossible."

In my opinion, the heaven was created but I am not sure how; probably
in the form of smoke. The reason for that is the Almighty's saying, "Then He
directed Himself to the heaven" that proves its existence. He did not follow that
by saying He created seven heavens but rather "and made them." If you read
the Quran, you surely came across these verses, "**Who created you, proportioned you, and balanced you?**" [Al-ʾInfitār:7], and the other saying of

the Exalted and Glorified, **"Who created and proportioned"** [Al-A'lā:2]. Hence, the creation took place first. If you were to understand the phases and the differences in these aforementioned verses, you would definitely understand them here in this context too. You would also realize that the heaven has sundry meanings apart from the common linguistic one that refers to that which is high above. Even more, the linguistic meaning of "highness" does not make sense unless juxtaposed with "the ground." And Allah knows best.

In this regard, I would like to note that the question of the creation of the heavens and the Earth and which one was created first is one that is open to argument and disagreement. Before detailing the disagreements, it is worth mentioning that many of the adversaries and foes of Islam question the authenticity of the Glorious Quran and regard it with contempt because the explicit meaning of the Quran indicates that the Earth was created before the heavens. This obviously goes against the conclusions of modern science that stipulates that the Earth is relatively new compared to the heavens, galaxies, and stars. Therefore, they raise suspicion in the hearts of those weak Muslims who blindly take for granted everything the West has come up with. Fortunately, some young Muslims, who are very enthusiastic about defending their faith, undertook the responsibility of refuting such misconceptions in the hope that they could do so using the mechanism of change, that is, changing the explicit meanings. In so doing, they sought to make the Quran compatible with modern science. In part, they managed, albeit for themselves, to turn the misconception into a scientific miracle in the Quran. Nevertheless, the real issue is not what the adversary nor the zealot think. The issue is that while the former took a flawed point of departure and offered a sound reading, the latter took a sound point of departure but offered a flawed reading. I beseech Allah to help us overcome the sheer ignorance everywhere around us that rears its ugly head in every direction. The moment you cut one of its heads, numerous new heads pop up over and over again, just like the well-known mythical hydra in Greek mythology. Allah is the best Helper.

As for the one who believes in the explicit meaning of the Quran and, at the same time, believes that Allah is the Truth and His words are true, this will not be hard for him, and he would even wonder, "What is the issue of creating the earth before the heavens? Do humans build their houses top down starting from the ceiling first or from the bottom up?" This is a plain example that a peasant on his farm understands and so does a baker in his bakery, a scientist in his laboratory, and the monk in his sanctuary.

On the whole, let's look at the positions of Quran exegetes on this issue. Some commentators argued that the Earth was created first while others held that the heaven was created first. Yet a third group opined that the mass of earth was created first, then the heavens were created, and after that the Lord leveled the Earth. As such, they reconciled between these verses that appear contradictory particularly for those whose comprehension and knowledge of the language are shaky. This very argument is the one the bulk of contemporary commentators subscribed to. One to whom this argument was ascribed to is Ibn 'Abbās who is reported to have said, "Allah created the heavens and the Earth. After creating the heavens and before creating the substances of the Earth, He then put the mountains in place. This means establishing the essentials of life on Earth because the substances and the living organism will not survive unless there is day and night. This is noted in His saying, '**And after that He spread the Earth**' [Al-Nazi'āt:30]. Did you not hear of His statement right after that? '**He extracted from it its water and its pasture**' [Al-Nazi'āt:31]."

'Abdullāh ibn Sallām maintained, "Allah started the creation on Sunday. He created the two earths on Sunday and Monday. Then He created the substances and the mountains on Tuesday and Wednesday. Then He created the heavens on Thursday and Friday and completed them before the last hour of Friday. In this hour, He created Adam in a hurry. Therefore, this is the hour in which the Day of Resurrection takes place." This narration that is arguably attributed to the Companion 'Abdullāh ibn Sallām indicates the creation of earth occurred before the creation of the heavens. Mujāhid remarked, "The creation of earth took place before that of the heaven. When He created the Earth, smoke arose from it. This is when '**He directed Himself to the heaven, [His being above all creation], and made them seven heavens, and He is Knowing of all things**' [Al-Baqara:29]." He added "heavens one on top of the other and seven earths one below the other." Al-Ḥassan related that, "Allah created the Earth in the site of Jerusalem in the form of a rock that has smoke all over it. He then raised the smoke and made it into the heavens. He then kept the rock in its place and spread it out into the Earth. This is what is meant by '**were a joined entity**' [Al-Anbiyā':30], i.e. they were attached together."

While commenting on this very verse, Ibn Jarīr al-Ṭabarī cited some accounts, and I will relate one of them, here. "Muḥammad ibn Ḥumayd narrated to us on the authority of Salama ibn al-Faḍl on the authority of Muḥammad ibn Isḥāq who reported, 'The first things Allah, Exalted and Glorified, created were the light and darkness, Then He set them apart making the darkness a dark, black night whereas the light a bright light day. Then He formed the seven heavens from smoke. It is said, and Allah knows best, that He created them from the smoke of water until they stood above high on their own. In the heaven, He dimmed its night, and brought forth its daylight. Henceforth, the day and night came into being. Yet, there was no sun, moon or stars in it. Then He spread the Earth, set the mountains firmly in it, determined its substances and scattered of whatever He wanted of creatures throughout. When He has done so on earth in four days, He turned toward the heaven that was smoke then as He stated. He structured them into seven, adorned the lowest sky with its sun, moon and stars and assigned to each heaven its command. He completed creating them in two days. As such, He completed the creation of the heavens and the Earth in six days.

"'On the seventh day, he settled on His heavens and then called upon the heavens and the Earth '**Then He directed Himself to the heaven while it was smoke and said to it and to the Earth, 'Come [into being], willingly or by compulsion.' They said, 'We have come willingly**' [Fuṣṣilat:11], i.e. to come to the form He has willed for them willingly or unwillingly. In response, they said, '**We have come willingly**' [Fuṣṣilat:11]' He added, 'Ibn Isḥāq maintained that Allah, Exalted and Glorified, turn to the heaven after creating the Earth and that is therein and they were seven heavens of smoke and He formed them as He has described.'

"On commenting on this verse, Ibn Kathīr, may Allah have mercy on him, posited, 'This is an indication that Allah, Almighty, started with creating the Earth first and then made the heavens into seven. This is the custom of construction; you start from the bottom up.' Other commentators underscored the same point. Al-Zamakhsharī expounded, 'the mass of earth was created before the heaven.' Al-Fakhr al-Rāzī stated, 'He created the heaven after the Earth and He left no interval in between and He did not turn to anything else after the creation of earth.' Al-Rāzī also noted, 'The Almighty's statement, **'Say, 'Do you indeed disbelieve in He who created the Earth in two days and attribute to Him equals? That is the Lord of the worlds.' And He placed on the Earth firmly set mountains over its surface, and He blessed it and determined therein its [creature's] sustenance in four days without distinction—for [the information] of those who ask'** [Fuṣṣilat:9–10], means that the creation of the Earth took two days and the ordinance of substances took two more days. This is more akin to saying that 'The distance from Kufa to Medina takes twenty days and to Mecca thirty days,' meaning the entirety of the distance. Thus, He turned to the heaven in two more days, making the sum six days."

Al-Qurṭubī maintained, "This verse clearly shows that the Almighty created the Earth before the heaven as He asserted in the chapter of al-Sajda and in al-Nazi'āt, **'Are you a more difficult creation or is the heaven? Allah constructed it'** [Al-Nazi'āt:27], where He described its creation. Then He added, **'And after that He spread the Earth'** [Al-Nazi'āt:30]. Therefore, it seems as if the heaven was created prior to the Earth. Allah, Exalted and Glorified, said, **'[All] praise is [due] to Allah, who created the heavens and the Earth and made the darkness and the light'** [Al-An'ām:1]. Qatāda held the same view that the heaven was created first. Al-Ṭabarī reported that from him. Mujāhid and other commentators maintained, 'The Almighty caused the water on which His throne rested to cry out and turned it into earth. Then smoke arose from it and rose up high and He made it into the heaven. As such, the creation of the Earth preceded the creation of the heaven. After that, He turned to the heaven and proportioned it into seven heavens. Then He flattened the Earth because it was not flat when He first created it.' I would argue that the position of Qatāda can be detailed flawlessly, God willing, as follows: Allah, Almighty, first created the smoke of the heaven and then created the Earth. Next, He turned towards the heaven while it was smoke and He made it into seven heavens. After that, He levelled the Earth flat.

"The evidence that the smoke was created first before the Earth lies in what al-Suddī reported who narrated from Abū Mālik, in what Abū Ṣālih reported on the authority of Ibn 'Abbās, in what Murra ibn al-Hamdānī on the authority of Ibn Mas'ūd, and in what is related from some of the Companions of the Messenger of Allah (peace be upon him) regarding the Almighty's saying, '**It is He who created for you all of that which is on the Earth. Then He directed Himself to the heaven, [His being above all creation], and made them seven heavens, and He is Knowing of all things**' [Al-Baqara:29]. He said that the throne of Allah, Exalted and Glorified, was on the water and He did not create anything before the water. When He wanted to create the creation, He invoked smoke out of the water and the smoke rose up high above the water. When it looked lofty so he called it heaven. Then, He dried out the water and made it a single earth which He later split into seven earths in two days on Sunday and Monday. '**It is He who created for you all of that which is on the Earth. Then He directed Himself to the heaven, [His being above all creation], and made them seven heavens, and He is Knowing of all things**' [Al-Baqara:29]. Then, he placed the Earth on a whale. The whale is the *nūn* fish, which Allah, Exalted and Glorified, mentioned in the Quran in His statement, '**Nun. By the pen and what they inscribe**' [Al-Qalam:1]. The whale was in the water; the water was on a solid surface; the surface on the back of an angel; the angel on a rock; and the rock is up in the air. It is the rock that Luqmān described as not in the heaven nor on the Earth. When the whale moved disturbingly, the Earth was shaken.

"Therefore, he set the mountains in it so it became stable. The mountains brag about that over the Earth. This is what He referred to in the Almighty's saying, '**And He has cast into the Earth firmly set mountains, lest it shift with you, and [made] rivers and roads, that you may be guided**' [Al-Naḥl:15]. Then He created the mountains in it and ordained means of sustenance for its inhabitants, its trees and other essentials of life in two days on Tuesday and Wednesday as He declared, 'Say, Do you indeed disbelieve in He who created the Earth in two days and attribute to Him equals? That is the Lord of the worlds.' '**And He placed on the Earth firmly set mountains over its surface, and He blessed it and determined therein its [creatures']** sustenance in four days without distinction—for [the information] of those who ask' [Fuṣṣilat:9–10]. He says, for whoever wonders, such was the matter. 'Then He directed Himself to the heaven while it was smoke and said to it and to the Earth, 'Come [into being], willingly or by compulsion.' They said, 'We have come willingly'' [Fuṣṣilat:9–10]. Such smoke was the exhale of the water when it breathed out. Thus, He made it one heaven and then He split it up into seven heavens in two days on Thursday and Friday. Friday (*Yaum al-Jum'a*) is dubbed as such because it marked the congregation of all creatures in the heavens and earth together. 'And He completed them as seven heavens within two days and inspired in each heaven its command' [Fuṣṣilat:12]. He added, He created in each heaven its creatures from angels as well as such other creatures as seas and mountains of hail and things we know nothing about. Next, He adorned the lowest sky with planets, making it an adornment and bulwark that guard them from devils. When He completed the creation of all he willed, He then established Himself on the throne. This is when He relates, 'It is He who created the heavens and the Earth in six days and then established Himself above the Throne' [Al-Ḥadīd:4]. He further says, 'Have those who disbelieved not considered that the heavens and the Earth were a joined entity, and We separated them' [Al-Anbiyā':30]." In the end, al-Qurṭubī concluded, "Only Allah knows for sure what He had done. It is a matter open to disagreement and there is no room in it for speculative reasoning."

As for *Tafsīr al-Jalālayn*, it posits, "It is He who created for you all of that which is on the Earth" [Al-Baqara:29], i.e. the Earth and everything on it. 'All of that' in order to benefit from it and to contemplate. 'Then He directed Himself to the heaven' after the creation of the Earth." In his commentary, al-Khāzin maintained, "He created the Earth first. Then He turned to create the heaven. If one were to ask how to reconcile this with the other verse, 'And after that He spread the Earth' [Al-Nazi'āt:30], I would reply: spreading it out means making it flat. It is probable that Allah, Almighty, created the mass of the Earth but without leveling it. Then He created the heaven and afterwards flattened the mass of the Earth. If you were to argue back that this is problematic as well because Allah, Almighty, said, 'It is He who created for you all of that which is on the

Earth' [Al-Baqara:29], and this entails that creating all these things is not possible before spreading it and making it flat, I would say in response that there could be no chronological order in this regard. It could be mentioned merely to enumerate the favors. It is like when someone says to somebody else he does favors to: Did not I give you? Did not I raise your status? Did not I guard you? It could be possible that one of these favors preceded another. And Allah knows best."

The most coherent interpretation is that of Abū Ḥayyān in his commentary, "This verse merely reveals that the creation of everything is on Earth for us precedes making the heaven into seven. The preferable hypothesis is that the mass of the Earth was created before the heaven and that the heaven was created after that. Then, the Earth was spread out and made flat after the creation of the heaven. Accordingly, it is possible to reconcile all these verses."

Al-Shawkānī suggested a salient interpretation of this verse, "The Almighty's saying 'He then directed Himself to the Heaven' has been used as an evidence that the creation of the Earth preceded the creation of the heaven. The same goes for the verse in the chapter of al-Sajda. He also said in al-Nazi'āt, '**Are you a more difficult creation or is the heaven? Allah constructed it**' [al-Nazi'āt:27]. So He described its creation. Then He added, '**And after that He spread the Earth**' [al-Nazi'āt:30]. As such, it seems that the heaven was created before the Earth. The same applies to the Almighty's statement, '**[All] praise is [due] to Allah, who created the heavens and the Earth and made the darkness and the light**' [Al-An'ām:1]. It is said that the mass of the Earth was created before the heaven but spreading it out took place later."

A number of eminent scholars subscribed to this interpretation that is an acceptable reconciliation of these verses that we cannot help but to resort to. However, the creation of things on earth can only occur after spreading it out. The verse in question indicated that He created whatever on earth before creating the heaven. Importantly, this means that the contradiction still exists, and such aforementioned reconciliation of verses does not steer clear of it."

Al-Nasafī maintained in his commentary, "Allah, Almighty, said, '**Then He directed Himself to the heaven while it was smoke and said to it and to the Earth, 'Come [into being], willingly or by compulsion.' They said, 'We have come willingly'**'[Fuṣṣilat:11]. The verse means that

He went purposely to create the heaven right after He had created the Earth without intending to create anything else in between." He also added, "The creation of the mass of the Earth was before the creation of the heaven but leveling it flat occurred later." Likewise, the renowned exegete al-Tha'ālibī stated, "This verse entails that the Earth and all on it were created before the heaven and this is factually true. Then the Earth was spread out and flattened after the creation of the heaven. This way the meaning of these verses and those of in the chapters of al-Mu'min (Ghāfir) and al-Nazi'āt are consistent." It is reported in the commentary of al-Biqā'ī that, "the creation of the Earth and preparing it for what it meant for occurred before the creation of the heaven but spreading it out was the creation of the heaven."

As far as the 'Ibāḍī commentaries are concerned, Aṭfīsh comments, "This verse shows that the Earth and all on it were created before the heaven. That the Earth here means its orientation is clear as we have explained earlier. That the Earth here means the mass of the Earth is because creating the Earth and everything on it gives the impression that it was created before the heaven. What readily comes to mind is that He created all that is on earth and then made the Earth their setting. It does not mean that He created these things in the air and then made the Earth it is setting. The term *thumma* (then) here reveals that about the Earth because it is a term for separation and succession. As such, the interpretation of the Almighty's saying, '**And after that He spread the Earth**' [al-Nazi'āt:30] will not be problematic because it means after the creation of heaven because what took place after the creation of heavens is the spreading out of earth, namely making its surface flat. Allah, Exalted and Glorified, created the Earth spherical, then created seven heavens and then spread out the Earth and made its surface flat. If one were to argue that custom says otherwise and it is far-fetched to create all that is in it for us while it is spherical. In response, I would say, that person denies that given the gargantuan size of earth but being spherical does mean that things will not rest on it. Moreover, the referent in the Almighty's saying 'after that' could go back to refer to the formation of the heaven, i.e. forming it into seven heavens. Consequently, the mass of earth precedes the mass of the heaven. The formation of the heaven, which is to make it into seven heavens, precedes the formation of earth, which is to spread and level it out. Therefore, it is befitting to interpret 'heaven' in the verse 'He directed Himself to the heave' to mean a lofty mass that was there after the creation of earth and that was as seven-heaven thick. He spread out the Earth after separating that mass into seven layers."

After relating the interpretations of the commentators, he remarked, "The truth is what I have mentioned at the beginning that the Earth was before the heaven and that the levelling of the Earth was after the formation of the heaven in order to reconcile the meaning of all verses."

Lastly, the commentary of Ibn 'Arafa suggests an answer to this salient big question and I relate it here for your benefit. "It could be argued that the heavens and earth were created attached and then the Earth was created and leveled flat. Then, the heavens were separated and made seven heavens, and Allah knows best."

As you have seen, most commentators interpreted the explicit meanings of the Quran, arguing that the Earth was created before the heavens. That is why they disagreed on the interpretation of Allah's, Exalted and Glorified, saying **"And after that He spread the Earth"** [al-Nazi'āt:30]. Otherwise, if the literal meaning of the Quran were to show that the heavens were created before the Earth, they would not have to bother themselves to reconcile these verses and to clarify any apparent contradictions between them.

Know that these verses deal a heavy blow to the conclusions of science that are widely accepted today that the Earth came into being four and a half billion years ago and that the cosmos came into being around fourteen billion years ago. How could that be consistent with the explicit meaning of the Quran, the prophetic traditions, the accounts of the Companions, the Successors, and the scholars from all walks of life, let alone with logical and common sense?

Indeed, some of us fell in the trap of believing these conclusions without any scrutiny or knowledge but this happened before we knew the doctrines of our faith and the creed we believe in. As I always maintain, "Whoever does not believe in Allah and His words, he believes is something else, and this is a fact of life."

9

BEGINNING OF THE HEAVENS AND EARTH

ALLAH, EXALTED AND Glorified, said, "**Have those who disbelieved not considered that the heavens and the earth were a joined entity, and We separated them and made from water every living thing? Then will they not believe?**" [Al-Anbiyāʾ:30].

The scholars proffered sundry interpretations for this verse, with the majority of them holding the view that the heavens and the earth were one mass and Allah then split them and made them two separate entities. Others offered a different line of argument, insisting the heaven was bereft of rain and the Earth was bereft of fauna and flora. Then Allah provided the former and the latter with water. They corroborated their view with the later segment of the verse. A third group opined that the heavens were one single mass until Allah split it into seven heavens. The same applies to the Earth, which was one mass, and He split it into seven earths. The argument of al-Rāzī, which some people who assimilated in the Western society take pride in, sets forth that the first and the second argument carry much weight and that the heavens and earth were one single mass, and Allah, the Exalted and Glorified, separated them. Myriad verses of the Quran refer to such partition depicting Allah as the Originator of the heavens and the earth. By the same token, the Torah, the Jewish, and Christian written texts indicate that Allah has created a partition or a divider between the heaven and the earth.

Ibn ʿAbbās, al-Ḥassan, ʿAtāʾ, Al-Ḍaḥḥāk, and Qatada maintained, "It means that they were attached as one mass and Allah then separate them with air." Likewise, Kaʿb also maintained, "Allah created the heavens and the earth one top of one another. Then He created gaseous air in between, separating them from one another and splitting the heavens into seven heavens and the earth into seven as well." Another argument offered by Mujāhid, al-Suddī, Abū Ṣālih contends, "The heavens were amalgamated in one single layer and He separated them into seven heavens. Likewise, the earth was coalesced into one single layer and He split it into seven as well." Ibn Qutayba narrated a similar argument in his magnum opus, ʿUyūn al-Akhbār, on the authority of Ismaʿīl ibn Abī Khalid with regard to the statement of Allah, Exalted and Glorified, **"Have those who disbelieved not considered that the heavens and the earth were a joined entity, and We separated them"** [Al-Anbiyāʾ:30]. He stated, "The heaven was created as a lone mass and the earth as a lone mass, too. Then Allah split the former into seven heavens and the latter into seven earths. He created the upper earth, made human beings and jinn its inhabitants, made streams flow through it, made flora grow on it, caused seas to flow through it and made its sky as a ceiling. Its width is width is as long as five hundred years' walk. Then He created the second one in the same breadth and width and made it inhibited by certain people, whose mouths are more akin to those of dogs, their hands are like those of humans, and their ears like those of cows and their hair is like that of sheep. When the Day of Resurrection takes place, this earth will throw those people to Gog and Magog. This earth is called al-Dakmāʾ (the dark earth).

"Then He created the third earth whose width is as long as five hundred years' walk and from which air goes to the earth. He created in the fourth earth darkness and scorpions for the dwellers of the hellfire such as the black gigantic scorpions, which have tails like those of towering horses and that eat one another. They are let to attack the children of Adam. Likewise, Allah created the fifth one in the same breadth, width and height. It has chains, shackles and fetters for the dwellers of the hellfire. Then Allah created the sixth earth, which is dubbed ʿMādʾ and which has extremely dark black stones from which the clay of Adam was created. These stones will be resurrected on the Day of

Resurrection as colossal towering mountains. Made of brimstone, they will be hung around the necks of the disbelievers, set ablaze then burn their faces and their hands. This is in reference to the saying Allah, Almighty, '**whose fuel is people and stones'** [Al-Baqara:24]. Then Allah created the seventh earth, which is called *'Arība,* and it has a hellfire that has two doors; one is called *Sajīn* and the other *al-Falaq. Sajīn* is open and it is the terminus point of the disbelievers and through which the people of the spread table and the community of Pharaoh are brought before the fire. As for *al-Falaq,* it is closed and will only be opened on the Day of Judgment."

As it was explained earlier in the chapter of al-Baqara that there are seven earths, with the interval between each being as long as five hundred years' walk. In the last part of the chapter of *al-Ṭalāq,* we shall elaborate more on this. A third argument espoused by 'Ikrīma, Ibn 'Aṭiyya, Ibn Zayd, and Ibn 'Abbās, as was stated by al-Mahdawi, contends that, "The heavens were bereft of rain and the earth was barren and bereft of flora. Then Allah fissured the heaven with rain and clefted the earth with plants. This meaning corresponds the statement of the Exalted and Glorified, '**By the sky which returns [rain] And [by] the earth which cracks open'** [Al-Ṭāriq:11–12]." This view was espoused by al-Ṭabarī, given that the following verse reads, "**and [We] made from water every living thing? Then will they not believe?**" [Al-Anbiyā':30]. I add that this view is corroborated both by means of observation and watching, and that is why this meaning has been reiterated in a number of verses as a clear manifestation of His Magnificent Omnipotence and of resurrection and reward and punishment.

It is said:

Easy is it for them when angered to eschew animosity or to inflict it.

To cause cracked-open things to be solid or to make the solid ones to crack open, and to undermine matters or to conclude them.

The statement of Allah, Almighty, "**and [We] made from water every living thing?**" [Al-Anbiyā':30], is open to three interpretations. First, He created everything from water, as argued by Qatada. Second, He preserved the life of everything with water. Third, He made every living thing of backbone (Solb) water. This is the view of Qutrub, and "made" here means "created."

In his sound hadith collection, Abū Ḥātim al-Bastī narrated on the authority of Abū Hurayra who stated, "I said, 'O Messenger of Allah, whenever I see
you, you make my day and you make me feel at ease. Kindly inform me about
everything.' He replied, 'Everything was created from water,'" as per the full
version of this hadith. Abū Ḥātim expounded on Abū Hurayra's words saying
that "Kindly inform me about everything" refers to everything that was created from water. This is substantiated by the Prophet's reply to that request by
saying, "Everything was created from water," even if it is not created. This is a
further argument in addition to what we have mentioned earlier that the heavens and earth are one solid mass. It is also argued that "everything" could mean
"some," as it is the case in this verse, **"Indeed, I found [there] a woman
ruling them, and she has been given of all things, and she has a great
throne"** [Al-Naml:23], and the other verse, "Destroying everything by command of its Lord" [Al-Aḥqāf:25]. The sound view, however, is that it means
everything by and large, as the Prophet (peace be upon him) said, everything
was created from water, and Allah knows best.

His saying **"Then will they not believe**?" [Al-Anbiyā':30] questions how
they would still not believe, in spite of what they observe and see. Evidently,
this maker per se cannot be created.

Allah, Almighty, maintains to highlight His Absolute Omnipotence, His
magnificent sovereignty, and His hegemony over His creations, **"Have those
who disbelieved not considered that"** [Al-Anbiyā':30]. He questioned
how those who deny His supreme sovereignty and associate partners with Him
still cannot realize that Allah is the supreme Creator and the sole Sustainer of
affairs. How can they dare to associate partners with Him or to worship them
besides Him? Have they not known that the heavens and the Earth were in the
beginning one single mass attached to one another and layered one on top of
the other, and then He separated each one from the other, making the heaven
into seven heavens and rendering the earth into seven as well? He separated the
heaven and the earth with air. Thus, the sky was made to rain, and the Earth
was made to grow plants. Therefore, He added, **"and [We] made from water
every living thing?"** [Al-Anbiyā':30], i.e. while they see with their own eyes,
the creatures being created constantly. All these are indications of the existence

of the Maker, the Sovereign, the Omnipotent, and the Most Capable of every-thing. For His part, everything indicates He is the Absolute One.

Abū Sufyān al-Thawrī related on the authority of his father, who related from ʿIkrīma, who said, "Ibn ʿAbbās was asked about what was first, the night or the day." He replied, "Have you not seen that when the heavens and earth were one attached mass, only darkness existed between them? This is so you know that the night preceded the day in existence."

Ibn Abū Ḥātim narrated from his father on the authority of Ibrāhīm ibn Ḥamza who narrated from Ḥātim on the authority of Ḥamza ibn Abū Muḥammad from ʿAbdullāh ibn Dinār from Ibn ʿUmar that a man asked him how the heavens and the earth were one attached mass and Allah separated them. He advised the man to go ask the old man and then to return to him and tell what that old man had told him. The man went to Ibn ʿAbbās and asked. Thereupon, Ibn ʿAbbās replied that indeed the heavens was bereft of rain and the earth was barren. When He created inhabitants for the earth, He fissured the earth with plants and the sky with rain. Then, the man came back to Ibn ʿUmar and related to him Ibn ʿAbbās' reply. Thereupon, Ibn ʿUmar remarked, "Now I know that Ibn ʿAbbās was granted so much knowledge about the Qurʾān. He is right and such was the way the heavens and earth were created." Ibn ʿUmar further reiterated that he used to take issue with Ibn ʿAbbās' endeavors to interpret the Qurʾān, but now he rested assured, given that Ibn ʿAbbās was granted so much knowledge about the Qurʾān. ʿAtiyya al-ʿŪfī said that one was rainless and then it was made to rain whereas the other was barren and plantless and it was made to grow plants.

Ismāʿīl ibn Abū Khalid said that he asked Abū Ṣāliḥ al-Hanafī about the interpretation of **"the heavens and the earth were a joined entity, and We separated them"** [Al-Anbiyāʾ:30]. The latter replied, "The heaven was just one and He split it into seven heavens. Likewise, the earth was one and He split it into seven earths. Mujāhid offer a similar line of argument, adding the heaven and the earth were not attached."

Saʿīd ibn Jubayr maintained, "Rather, the heaven and the earth were attached but He separated the sky from the earth. That this is the separation of them that Allah referred to in His Scripture." Al-Ḥassan and Qatada stated that they were attached, and He separated them with the air, and that His saying, **"and [We] made from water every living thing?"** [Al-Anbiyāʾ:30] means the essence of all living things originates from it. Ibn Abū Ḥātim reported on the authority of his father from Abū al-Jamāhir who narrated from Saʿīd ibn Bashīr who related from Qatada from Abū Maymūna from Abū Hurayra who said, "'O prophet of Allah! When I see you, you made my day and put me at ease, so tell us about everything.' He replied, 'Everything was created from water.'"

Imām Aḥmad recorded that Yazīd who narrated on the authority of Hamām from Qatada from Abū Maymūna from Abū Hurayra who said, "I said, 'O messenger of Allah! When I see you, you made my day and put me at ease, so tell us about everything.' He replied, 'Everything was created from water.' Then went on to ask, 'tell me about something if I were to do, I will be admitted into para-dise.' He replied, 'exchange greetings of peace (i.e., say: As-Salamu ʿAlaikum to one another), feed people, strengthen the ties of kinship, and be in prayer when others are asleep, you will enter paradise in peace.'"

He also recorded it on the authority of ʿAdel-Ṣamad and ʿAffān and Bahz from Hamām. Aḥmad is the only one who narrated this hadith and this chain of transmitters adhere to the authenticity guidelines stipulated by the collectors of the two-sound hadith. Yet, Abū Maymūna is a transmitter who meets the condition set by the collectors of the *Sunnan* hadith collections, and his name is Silīm. Al-Tirmizī deems his narrations sound. Saʿīd ibn Abū ʿAūba narrated it from Qatada as a *Mursal* hadith related by a successor or a companion who did not hear it from the Prophet. Allah knows best."

The statement of Allah, Almighty, **"And We placed within the earth firmly set mountains"** [Al-Anbiyāʾ:30] means He put firm mountains to sta-bilize the earth, lest it should sway and move under the people. If it were to shake with them, it would be difficult for them to live stably on it, given the fact that all of it is submerged in water except one fourth of it, which is visible for air and the sun, and in order for the inhabitants of that part to see the sky and observe what it has to offer of myriad spectacular signs, attestations, and manifestations. Therefore, He added, "lest it should shift with them," that is, so that it may not sway under them. The statement of Allah, Almighty, "and We made therein [mountain] passes [as] roads" means that We made broad pathways throughout it to serve as roads from one country to another and from one region to another as it is evidently observed on Earth where mountains stands as barriers between such-and-such country. Thus, Allah made a pathway through it to enable people to cross from here to there. He thus concluded the verse with "that they might be guided."

As for Allah's saying, "And We made the sky a protected ceiling" [Al-Anbiyā':32], it refers to a ceiling on the Earth, similar to a dome over it. This is more akin to His other sayings, **"And the heaven We constructed with strength, and indeed, We are [its] expander"** [Al-Zāriīat:47], "And [by] the sky and He who constructed it" [Al-Shams:5], and **"Have they not looked at the heaven above them—how We structured it and adorned it and [how] it has no rifts?"** [Qāf:6]. The ceiling is the structure of the dome as the Messenger of Allah (Peace be upon him) said, "Islam is built upon five," namely five pillars, and this is especially true for tents as it is known by Arabs. His saying "protected" means it is high, guarded, and unreachable. Mujāhid, however, said it means towering high.

Ibn Abū Ḥātim stated that ʿAlī ibn al-Ḥussein related to him on the authority of Aḥmad ibn ʿAbdal-Rahmān al-Dashtakī who reported from his father who narrated from his father from Ashʿath, that is, Isḥāq al-Qommī, from Jaʿfar ibn Abū al-Mughira from Saʿīd ibn Jubayr from Ibn ʿAbbās who said a man asked, "O Messenger of Allah, what is this sky?" He replied, "It is waves that have been lifted for you." This hadith is conveyed by only one narrator in one phase of its chain of transmission.

Allah's saying, "but they, from its signs, are turning away" [Al-Anbiyā':32] is similar in meaning to His other saying, **"And how many a sign within the heavens and earth do they pass over while they, therefrom, are turning away"** {Yusuf:105], that is to say they never ponder over what Allah has created in them, including their vast width and lofty height, as well as what they were decorated with such as the static celestial planets and the dynamic ones that revolve day and night in them, not to mention the sun that covers the whole universe in a day and night. It moves such a long distance that only Allah knows of because He is the one Who created it and made it move as such.

Ibn Abū al-Dunya, may Allah have mercy on him, maintained in his book *Al-tafakkur wal-I'tibār*, that, "Some worshippers among the people of Israel worshipped for thirty years. When a man amongst them were to worship for thirty years, a cloud would shade him and that person will not be able to see what happens to others. So he would complain to his/her mother who would tell him: Probably you have committed a sin while worshipping? He replies in the negative. So she would say: Maybe you only wanted to. He once again replies in the negative saying that he never wanted to do so. Then she says: Maybe you have looked up and then down but failed to ponder and contemplate? He then replies in the affirmative, adding that he did that a lot. She then said: Whence you were inflicted."

Allah further adds, "And it is He who created the night and the day" [Al-Anbiyā':33], referring to the former with its darkness and serenity and the latter with its light and liveliness. He prolongs the former at a time and shortens the latter at another, and vice versa. By the same token, His saying **"And the sun and the moon"** means the former has its own light, orbit, timing, and movement while the latter has its own different light, orbit, movement, and timing, since **"all [heavenly bodies] in an orbit are swimming"** [Al-Anbiyā':33]. Ibn 'Abbās stated, "They revolve around in curved orbits in a circular motion." Likewise, Mujāhid said, "They float in curved circular orbits and they will not revolve around without such orbits. Similarly, the stars, the sun, the moon will not revolve unless in orbits." This corresponds to the statement of Allah, Almighty, **"[He is] the cleaver of daybreak and has made the night for rest and the sun and moon for calculation. That is the determination of the Exalted in Might, the Knowing"** [Al-An'ām:96].

Commentators have offered sundry interpretations of the terms *ratq* (attached mass) and *fatq* (separation) and the following are their views:

The first view is that of al-Ḥassan, Qatada, and Sa'īd ibn Jubayr and the narration of Ikrīma from Ibn 'Abbās (may Allah be pleased with them all). They contend that they were one attached mass, and then Allah separated them, lifting the sky to where it now, and stabilizing the earth. This view entails that the creation of the Earth preceded the creation of the heaven, because when Allah separated them, He left the earth where it was and lifted the upper parts constituting the heavens. Ka'b remarked, "Allah created the heavens and earth attached and then He created air in between them and thus separated them."

The second view is that of Abū Ṣāliḥ and Mujāhid. They opine that the heavens were attached to one another, and they were then split into seven heavens. The same applies to the Earth.

The third view is that of al-Ḥassan, Ibn 'Abbās, and the majority of commentators. They argue that the heavens and earth were solid and flat, and then Allah vented the former with rain and creviced the latter with plants and trees. This is similar to the other statement of Allah, Almighty, **"By the sky which returns [rain] And [by] the earth which cracks open"** [Al-Ṭāriq:11–12]. Scholars have argued that this view carries more weight because Allah, Almighty, stated in the same verse, **"and made from water every living thing?"** [Al-Anbiyā':30]. Importantly, this means the water has something to do with what was mentioned earlier in the verse. Thus, this meaning exquisitely substantiates this aforementioned view. Yet, one might argue that this view is far-fetched, given that water does not descend from all heavens but rather from the lowest one, that is, the sky. In response, I would say that He used the plural because every layer of them is called a sky, per se. This is a common Arabic usage. Importantly, you know that, as per this interpretation, the term "*yarā*" in this verse could mean "see."

The fourth view is that of Abū Muslim al-Aṣfhānī. He argues that the term *Ratq* could mean creating and bringing into being, as in Allah's saying, **"[He is] Creator of the heavens and the Earth"** [Al-Shūra:11], and His other saying, **"He said, '[No], rather, your Lord is the Lord of the heavens and the earth who created them'"** [Al-Anbiyā':56]. For creation, He used the term *Fatq*, and for their condition before creation, He used the term *Ratq*. In response, I argue that this is evidenced by the fact that non-existence is a negation of everything with no dissimilar beings or different entities. If these truth beings existed, then during creation and development, some would become distinct from others and some would separate from others. This way, it is befitting to make the term *ratf* a metaphor for non-existence and the term *fatq* a metaphor for creation.

The fifth view contends that the night preceded the day, for Allah, Almighty, said, **"And a sign for them is the night. We remove from it [the light** of] day" [Yāsīn:37]. The heavens and the Earth were dark, so Allah, Almighty, split them by the creation of the bright daylight.

If one were to wonder which view conforms to the explicit meaning of the Quran the most, we respond that the explicit meaning entails that the sky is as it stands and the Earth as it stands were once attached and this necessitates that they were created then. Since *ratf* is the antonym of *fatq*, the former means being attached as one mass while the latter means separating them. As such, the fourth and fifth views are implausible, whereas the first view becomes the most plausible argument, followed by the second, the view that contends that the layers of the heavens and the earth were attached and then Allah split them into seven separate ones. Then comes the third view that argues they were solid with no fissures or cracks, and He cracked the sky with rain and the earth with plants. Al-Zamakhsharī concluded that "this means the sky and the earth were jointly attached with no space whatsoever in between."

10

SIX DAYS

ALLAH, ALMIGHTY, SAID, "**Indeed, your Lord is Allah, who created the heavens and earth in six days and then established Himself above the Throne. He covers the night with the day, [another night] chasing it rapidly; and [He created] the sun, the moon, and the stars, subjected by His command. Unquestionably, His is the creation and the command; blessed is Allah, Lord of the worlds**" [al-A'rāf:54]. Along the same lines, Allah, Almighty, said, "**Indeed, your Lord is Allah, who created the heavens and the earth in six days and then established Himself above the Throne, arranging the matter [of His creation]. There is no intercessor except after His permission. That is Allah, your Lord, so worship Him. Then will you not remember?**" [Yūnus:3]. In the chapter of al-Furqān, He, Almighty, said, "**He who created the heavens and the earth and what is between them in six days and then established Himself above the Throne—the Most Merciful, so ask about Him one well informed**" [Al-Furqān:59].

I will address several issues here, beginning with the issue of days. Allah, Exalted and Glorified, informed us that He created the heavens and the Earth in six days, but the modern scientific theories completely deny the creation of the heavens and the Earth in six human days while maintaining that the creation process extended beyond billions of years. Believers of modern science categorically refuse the statement of the creation of the heavens and the Earth in just six days. In pursuit of pleasing the scientific society, many contemporary scholars, thinkers, preachers, and educated men interpret the days as non-human days, but instead days of Allah, Exalted and Glorified. They maintain that only Allah, Almighty, knows the true nature of those days, and perhaps they are equal to billions of years, and Allah knows best.

There is no doubt about that but let me remind you of some issues. Allah, Exalted and Glorified, said, **"And they urge you to hasten the punishment. But Allah will never fail in His promise. And indeed, a day with your Lord is like a thousand years of those which you count"** [Al-Haj:47]. In Another verse, Allah said, **"The angels and the Spirit will ascend to Him during a Day the extent of which is fifty thousand years"** [Al-Ma'ārij:4]. Either one of the thousand years or the fifty thousand years will be at odds with the astronomical figures introduced by the modern science through its theories. They consider it impossible, or even comical, to maintain that Allah, Exalted and Glorified, created the heavens in the above mentioned period of time. In contrast, it is entirely logical for the believers in the shape of Earth Allah described to believe the six days were in human time.

Not only that but if you tracked down all the Quranic verses where Allah, Exalted and Glorified, created the heavens and the Earth in six days, you will find that some of those verses preceded, in revelation, the verses maintaining that a day is equal a thousand of fifty thousand years. Such an issue is inescapable by those who determine the Meccan or Medinan nature of some chapters and verses. If we were to consider some of the Meccan chapters and verses, we will find that some of them maintain Allah created the heavens and the Earth in six days.

The point I am trying to illustrate is that if we assume the case that those days were in non-human time, as proposed by the contemporaries of today, would it be logical to say that Allah repeatedly revealed to His Prophet (peace and blessings be upon him) the mention of days without being understandable by people what He exactly meant by those days? Furthermore, why had not anyone from the disbelievers or the People of the Book asked about the extent of those days? Is it not because it is clear that those days are the commonly known human days?

Assuming we follow the opinion maintaining that the Quran was revealed entirely at once or, at least, was revealed in the currently found order starting by the chapter of al-Fātiḥa, al-Baqarah, āl-Imrān, and so forth, there are many verses revealed before the ones mentioning that the day is equal a thousand or a fifty thousand years. Now, should you collect the verses under various themes or read the thematic/orderly exegesis (as preferred to be called as such by some contemporaries), you will arrive to the conclusion that all the verses mentioning that a day is equal a thousand or fifty thousand years are irrelevant to the topic of the creation of the heavens and the Earth. They have a completely different context.

Let us read another verse. Allah, Exalted and Glorified, said, "**It is Allah who created the heavens and the earth and whatever is between them in six days; then He established Himself above the Throne. You have not besides Him any protector or any intercessor; so will you not be reminded?**" [Al-Sajdah:4]. It contains one more mention of the six days. If they say that the days of Allah are different from our human days, the question posing itself would be, "Which verses were revealed firstly?" Don't you say that the Quran was revealed in parts? Since this is the case, this would mean the verse containing "**in a Day, the extent of which is a thousand years of those which you count**" [al-Sajdah:5] was revealed after some of the verses highlighting the six-days period. Nothing is left to say but to maintain that everyone should understand this, or the Jews would say, "No, this is wrong. It is actually six thousand years, but the books of Hadith had not recorded this."

Even the People of the Book, the exegetes of the Torah, and the Bible have generally considered that the days in which Allah created the heavens and the Earth are the six human days. Even some interpreters of the Bible like John Calvin and Martin Luther have emphasized this fact, though they were some-what figures of Enlightenment who cope with the evolution of science. In the case, this literal understanding had been mistaken in their Holy Bible. Perhaps Allah, Exalted and Glorified, would have corrected them clearly in the Quran, similar to His refutation of their claim that Jesus is the son of Allah or that Allah rested on the seventh day. Allah, Exalted and Glorified, said, "**And We did certainly create the heavens and earth and what is between them in**

six days, and there touched Us no weariness" [Qāf:38]. Reflect deeply on this subtle Quranic note, since it is very deep.

I would like to ask the educated and the Enlightened---as they prefer to call themselves or induce their followers to call them, "Don't you say that the Quran is graspable by even the farmers and laymen? Then, if a layman person were to read the verses stating Allah created the heavens and the Earth in six days, what would his understanding be? Wouldn't he think of the common days he knows? Do you think people, in general, would analyze the days as a special case?" No, that would not happen. I do not think anyone would answer in a way defying the literal meaning of the Quran. So, if this is the case, why would you deviate the literal meaning of the verse? Is it only for the purpose of reconciling them with some of the theories of the scientific society? Isn't this unacceptable in Islamic law? It is inappropriate to provide an alternative meaning to the literal one spelled out by the verse unless there is recognizable proof legitimizing this deviation to another hidden meaning. Why would you distort the meaning while the verses are in complete harmony with the cosmic outlook of the Quran?

Allah, Exalted and Glorified, said, "**Say, 'Do you indeed disbelieve in He who created the Earth in two days and attribute to Him equals? That is the Lord of the worlds'**" [Fuṣṣilat:9]. This is a Quranic fact that Allah, Exalted and Glorified, created the heavens in two days and the Earth in four. Have you ever inquired why? The recognizable scientific society of today maintain there are approximately a hundred thousand and million stars in a single galaxy, whose overall number is a hundred billion. The Earth, in comparison to this spacious cosmos, is equal to a tiny particle. My question is, "Why would Allah create the heavens in two days while spending four in the creation of the Earth, particularly when stars outnumber the grain of sand on Earth?" True, I completely understand that Allah has a rationale behind everything and that He does what He wills. However, have we ever tried to uncover some of the rationales spread throughout the Divine signs? In the Quran, Allah said, "**and a paradise as wide as the heavens and earth**" [āl-Imrān:133]. What is size of the Earth in comparison to the size of the heavens, as they claim today? In addition, what is the purpose of setting it in comparison with the paradise? Is this the only verse capable of being understood by people in a language they understand while stripping all the other Quranic verses their meanings? Have you ever seen someone showing off his latest model car by comparing it with a kid's plastic car? Have you ever seen a Japanese man comparing the Muramasa sword with a wooden one? You will never see this. Then why are they pleased by this comparison when they speak about the paradise and the Earth?

I hope they do not escape answering by saying that Allah has a rationale behind everything. They should have said that when they faced the dust thrown by the scientific society. Any one person with an understating of how comparison works would realize the gigantic size of the Earth to the extent that it can be placed onto one side of the balance while the heavens mentioned in the verse are placed on the other side.

11

WIDTH OF THE HEAVENS AND EARTH

"AND HASTEN TO **forgiveness from your Lord and a garden as wide as the heavens and earth, prepared for the righteous"** [Āl-'Imrān:133].

What is the size of the Earth as per those who espouse the theories of modern science? How can they imagine the Earth as a tiny, insignificant dot in this vast universe? Do they fathom the magnitude of paradise or not? Just juxtapose that with the view of the believer who believes the human existence is an Earth below and heavens high above as if one is in a house or a building. The size of this house or building is so colossal and the floor of the house is so enormous that nobody can fathom its magnitude. Then, someone told you that s/he will give you in lieu of your house in which you reside a paradise that is as wide as your house and it has a lot to offer. Definitely, there is a huge difference.

Meanwhile, we have to wonder, "Does the verse mean that its size is as vast as that of the heavens and that of the Earth? And what is the width of the Earth? Or does He mean it is as wide as both the heavens and the Earth combined, just as one would say, 'My house is as wide as yours,' or as you would say, 'The width of my house is as vast as that of your ceiling and the insignificant flower vase you have in your house?' Just look at the difference and you will understand what I mean.

We have an untold volume of narrations that are either ascribed to the Prophet (Peace be upon him) or to his companions. These accounts shed light on the way of thinking of people back then. For instance, in his commentary, Ibn Jarīr al-Ṭabarī maintained, "It was reported that the messenger of Allah (Peace be upon him) was asked: Since this paradise as wide as the heavens and earth, where is the hellfire? He replied, 'When the daylight alternates, where does the night go?'"

In what follows, we shall relate some narrations that were reported from the messenger of Allah (Peace be upon him) in that regard. On the authority of Yaʿlā ibn Murra who said, "I ran into al-Tannūkhī, the messenger of Hercules, to the messenger of Allah (Peace be upon him) in Homs when he was a housebound old man and he said, 'I come to the messenger of Allah (Peace be upon him) with the message of Hercules and he gave the parchment to a man on his left. I then said, 'Who can read amongst you?' They replied, 'Muʿawiya.' He then came saying, 'You called me to a paradise that is as wide as the heavens and earth that has been made ready for the righteous. If so, where is the hellfire?' Thereupon, the messenger of Allah (Peace be upon him) replied, 'Glory be to Allah. Where does the night go when the daylight appears?'"

Ṭāriq ibn Shehāb reported that some Jews asked ʿUmar ibn al-Khaṭṭāb about the paradise whose width is as vast as the heavens and earth and, if so, where the hellfire is. In response, he said, "Have you thought about where the night goes when the daylight appears?" They replied, "By God, we have found a statement more akin to that in the Torah."

Ṭāriq ibn Shehāb also reported that three people from Najrān came to ʿUmar while he was with his companions and asked him, "Have you known about '**a garden as wide as the heavens and earth**' and if so, where the hellfire will be?" The people in his company were taken aback. ʿUmar said in response, "Have you thought about where the night goes when the daylight appears, and where the daylight goes when the night alternates?" They then replied, "We have found a similar statement in the Torah."

Ṭāriq ibn Shehāb further reported that a Jew came to ʿUmar and wondered, "You say, '**a garden as wide as the heavens and earth**,' so where is the hellfire?" Umar said in response, "Have you thought about where the daylight goes when the night alternates, and where the night goes when the daylight appears?" The man then replied, "We have a similar statement in the Torah." His companion then remarked, "Why have you asked him then?" He said to his friend, "Leave him alone, for he believes in them both."

Yazīd ibn al-Aṣamm narrated that a man from the People of the Book came to Ibn ʿAbbās and asked him, "You say, '**a garden as wide as the heavens and earth,**' so where is the hellfire?" Ibn ʿAbbās said in response, "Have you thought about where the daylight goes when the night alternates, and where the night goes when the daylight appears?"

I even asked myself why Allah, Exalted and Glorified, said that it is as wide as the heavens and the Earth or its width is as vast as the sky and the Earth as He, Almighty, said elsewhere in the other verse, **"Race toward forgiveness from your Lord and a Garden whose width is like the width of the sky and the earth, prepared for those who believed in Allah and His messengers. That is the bounty of Allah which He gives to whom He wills, and Allah is the possessor of great bounty"** [Al-Ḥadīd:21]. I have repeatedly reflected on why He mentioned the width and not the length. I even went as far as telling myself that if the sky is a ceiling and the Earth is flat, then that means the length is far greater than the width, as we normally see and observe in buildings with the naked eye.

Then I thought about another thing, namely the equal length of the width of both the heavens and the earth from a certain point. Thus, the width of the sky from its lowest point to earth is equal to the width of the earth from the closest point to the sky. As a result, they are equal in terms of width. As far as the length is concerned, only Allah knows it. It is impossible for me or anyone to guess the length of the heavens or the depth of the earth. He declared that, **"whose width is like the width of the sky and the earth,"** which is like saying its width is tantamount to the limited space of existence of humans.

Another salient matter that requires thorough contemplation, deep thinking, unbounded imagination, and stronger relevance came across my mind. What if paradise exists in the sky on the seven heavens, and what if the hellfire exists underneath the Earth or in one of its corners, or around it, beyond the boundaries of the Earth or its circumferences and diameters? What if such was the case? In such a case, we find that the paradise could be as wide as the heavens and the Earth but may not be as long as they are. As such, the ceiling of paradise is the throne and things around it, as it was reported in some accounts attributed to the messenger of Allah (Peace be upon him).

If such were the case, the rationale of why He mentioned the width instead of the length would be clear. One of the hypothetical amazing suppositions is the notion of what if paradise were to substitute the heavens and earth in the Hereafter? Take a deep breath and recite the statement of Allah, Exalted and Glorified, **"And they will say, 'Praise to Allah, who has fulfilled for us His promise and made us inherit the earth [so] we may settle in Paradise wherever we will. And excellent is the reward of [righteous] workers'"** [Al-Zumar:74]. This verse proffers sundry indications for people of sound understanding who make cogent connections between things and thus open for themselves new horizons to better understand such meanings of the Quran that is granted to humans.

If we were to assume that the heavens and the Earth were a mere bubble in the universe, an oval object in space or something like that, how would it be possible for paradise to substitute them? This begs the question, "What had the people back then understood when these verses were sent down? Had they thought of the Earth as a dot in the space that is tinier than an atom, or had the thought of it as a colossal body that matches the heavens?" Definitely, the latter view carries more weight. If it became to us that this was true and that the Quran was sent down in a language very accessible to the sane people amongst them, be they of those who have a thorough understanding and mediocre comprehension capabilities. The recorded narrations we have provide no accounts of questions that might conjure up any doubt regarding the shape of the Earth. Hence, we conclude that the earth is indeed flat.

What I have elaborated on earlier is based on an imaginary, hypothetical supposition. Yet, from a rational point of view, I would argue—and Allah knows best—that the difference is that there is another verse in the chapter of Āl-'Imrān that reads, **"And hasten to forgiveness from your Lord and a garden as wide as the heavens and earth, prepared for the righteous"** [Āl-'Imrān:133]. The reason why He mentioned the heavens in that context is that because it is prepared for the righteous people. The term 'heavens' here is more specific than the sky which only means that which stands high above us. The aforementioned verse in the chapter of Āl-ʿImrān addresses the believers, particularly those who have believed in Allah and His messenger Muḥammad (peace be upon him) and hold higher ranks of belief. He described them to us as those **"Who spend [in the cause of Allah] during ease and hardship and who restrain anger and who pardon the people—and Allah loves the doers of good; And those who, when they commit an immorality or wrong themselves [by transgression], remember Allah and seek forgiveness for their sins—and who can forgive sins except Allah?—and [who] do not persist in what they have done while they know. Those—their reward is forgiveness from their Lord and gardens beneath which rivers flow [in Paradise], wherein they will abide eternally; and excellent is the reward of the [righteous] workers"** [Āl-ʿImrān:134–136]. He said, **"prepared for the righteous,"** and the righteous individual is the person who went beyond believing in Allah and His messengers. S/he is in a more sublime position and in a higher rank of faith. As for the verse in the chapter of al-Ḥadīd, He said, **"Race toward forgiveness from your Lord and a Garden whose width is like the width of the sky and the earth, prepared for those who believed in Allah and His messengers. That is the bounty of Allah which He gives to whom He wills, and Allah is the possessor of great bounty"** [Al-Ḥadīd:21]. Probably the address here is for human beings by and large. Therefore, He said, **"prepared for those who believed in Allah and His messengers."** and this is what is required from people. Moreover, one should ponder on how He concluded the verse by saying, **"That is the bounty of Allah which He gives to whom He wills, and Allah is the possessor of great bounty."** as if they did not do anything, and it is all a bounty from Allah. He did not list it as one of their characteristics as He did in the verse of the chapter of the Righteous, that is, Āl-ʿImrān, where He detailed many of the traits they have. Perhaps that is why He only mentioned the sky and the Earth, that is, what is beneath and what is above, without mentioning the heavens. Moreover, the singular heaven is broader than the plural heavens. Likewise, those believe in Allah and His messenger are broader and more general than the righteous who are specific.

On the whole, one should reflect on the fact that Allah said in the chapter of Āl-'Imrān **"And hasten to"** whereas he stated in the chapter of al-Ḥadīd, **"Race toward."** Why? I do not know for sure. Probably given that the righteous have no losers amongst them, and they are all good, they are the ones who hastened and the ones who did not. They are all on the safe side, if Allah wills so. On the other hand, a "race" has winners and losers. Given that the belief in Allah and His messengers is a very salient issue that gives rise to winning or losing, He used the term "race," so reflect on this superb Quranic gesture and give it its due heed in order for you to realize that the Quran cannot be the speech on anyone else but Allah. Glory be to Him, **"who has sent down upon His Servant the Book and has not made therein any deviance"** [Al-Kahf:1]. and the One Who said, **"And We send down of the Quran that which is healing and mercy for the believers, but it does not increase the wrongdoers except in loss"** [Al-Isrā':82].

12

FALL OF THE SKY UPON THE EARTH

ALLAH, ALMIGHTY, SAID, "**Do you not see that Allah has subjected to you whatever is on the earth and the ships which run through the sea by His command? And He restrains the sky from falling upon the earth, unless by His permission. Indeed Allah, to the people, is Kind and Merciful**" [al-Haj:65].

Allah, Almighty, said, "**Indeed, Allah holds the heavens and the earth, lest they cease. And if they should cease, no one could hold them [in place] after Him. Indeed, He is Forbearing and Forgiving**" [Fāṭir:41].

Why would Allah restrain the sky from falling upon the earth? Have you ever asked yourself this question? Remarkably, Allah created the sky with an inclination to fall. It is naturally set to fall, but He, Almighty, restrains from falling. Obviously, restraining something would not be possible unless it is liable to fall. One other issue is that the sky is something material, thus it is only rational that it may fall, which would annihilate people. It is not something immaterial or unphysical. It is rather a reinforced ceiling, as proven by the Quranic verses, like the ceiling of the house, which, if fallen, would kill everyone inside the house. The sky is similar on this regard, as I have mentioned in the previous chapters.

The scientific society maintains that the Earth is a tiny object roaming the space, and the sky is a space. The obvious question becomes: How could it be possible for the sky to fall on the Earth if we were to assume that the sky surrounding the Earth from all directions? Why had not Allah, Almighty, said that He restrains the sky from swallowing the Earth or the like? Instead, He said "fall," which signifies it is a drop from a higher to a lower position. How about the residents in the South of the Earth? From the perspective of the humans living there, it is possible it would fall. However, if we follow an external perspective for the Earth, the sky would have to be raised in order to fall on top of those who live under it in the South. The reality is that the sky does not surround the Earth but rather extends above it.

In his commentary, al-Shanqīty said, "In the following similar verses, '**And He restrains the sky from falling upon the earth;**' and, '**Indeed, Allah holds the heavens and the earth, lest they cease. And if they should cease, no one could hold them [in place] after Him;**' and, '**If We should will, We could cause the earth to swallow them or [could] let fall upon them fragments from the sky;**' and, '**And constructed above you seven strong [heavens];**' and, '**And the heaven We constructed with strength, and indeed, We are [its] expander;**' and, '**And We made the sky a protected ceiling,**' These verses clearly refute the claim made by non-Muslim atheists and their ignorant followers who claim Islam stating that the sky is a space, not a ceiling. Their claim is an expression of disbelief, atheism, and denial of the texts of the noble Quran; and Allah knows best."

Allah's, Almighty, saying, "**And We made the sky a protected ceiling**" is an explicit proof that the sky is above the Earth, not enveloping it. It is inconceivable to call the floor of your room or office a ceiling, nor do we call the walls surrounding you a ceiling. Only what is above you is called a ceiling. Allah, Exalted and Glorified, said, "**Those before them had already plotted, but Allah came at their building from the foundations, so the roof fell upon them from above them, and the punishment came to them from where they did not perceive**" [Al-Naḥl:26].

The fact that the sky is a ceiling is understandable and logical. But where is the ceiling of the Earth? Will you tell me it is above the millions of galaxies containing billions of stars wherein the Earth is nothing but a particle in comparison, assuming this was unknown to the generation who received the Quran first-hand? Indeed, their assumption is unreasonable. The Quran and its verses offer unequivocally indications that the sky is the ceiling of the Earth though they persist on the opposite.

Let us explore the statements made by exegetes when they commented on this verse. Al-Tabary said, "The sky is a ceiling for the Earth." In the commentary of al-Qurtuby, he said, "[The ceiling] is restrained from falling on the Earth, based on the saying of Allah, Almighty, '**And He restrains the sky from falling upon the earth, unless by His permission**.' It is said that it is restrained through stars against the devils, as maintained by al-Farrā' following the saying of Allah's, Almighty, '**And We have protected it from every devil expelled [from the mercy of Allah]**' [Al-Ḥijr:17]. Another opinion maintains that it is protected against destruction and it is far above anyone's schemes. Furthermore, it is said that it is raised and needless of a pillar."

In the commentary of ibn Kathīr, he said, "[The sky] is above the Earth like a dome, as Allah said, '**And the heaven We constructed with strength, and indeed, We are [its] expander;**' and, '**And [by] the sky and He who constructed it;**' and, '**Have they not looked at the heaven above them—how We structured it and adorned it and [how] it has no rifts?**' Construction here refers to the building of the dome, as the Messenger of Allah (peace and blessings be upon him) said, 'Islam is built on five pillars . . . ' meaning five posts, which is only conceivable in the setup of tents, typical of the habit of Arabs to make the ceiling high and protected against breach."

Al-Rāzī said, "The sky is called a ceiling because it serves the same purpose for the Earth like a ceiling for a house." He continued, "There are two opinions regarding the protected ceiling; one says that it is restrained from falling off like any regular ceiling, following Allah's saying, '**And He restrains the sky from falling upon the earth, unless by His permission;**' and, '**And of His signs is that the heaven and earth remain by His command. Then when He calls you with a [single] call from the earth, immediately you will come forth;**' and, '**Indeed, Allah holds the heavens and the earth, lest they cease**' [al-Rūm:25]; and, '**and their preservation tires Him not**' [al-Baqarah:255].

In the commentary of al-Biqāʿy, he said, "[The sky] is a ceiling for the Earth without the least difference between it and the regular ceilings, except in the harmful objects falling off regular ceiling. The ceiling sky, in contrast, is full of benefits. Most of what descends from it is indispensable for human life such as light, signs of navigation, and priceless ornaments. Normally speaking, regular ceilings, though small in size, need pillars for support, not to mention it is vulnerable for destruction and needy of regular maintenance. This ceiling sky, on the other hand, is the complete opposite despite its height and massive size. It is restrained from falling, by Allah's Power, and protected against devils by burning flames. Most people are averse to the signs [of the sky] such as big and small planets, wind and rain, and other countless signs. They prove the power of Allah over anything He wants, such as resurrection. They also prove Allah's greatness exemplified in His Oneness and other attributes of perfection and glorification. Most people deny those signs without reflecting on how they are operated and disposed into locations for rising and setting in a perfect order that indicate the equation governing all benefits."

I have not found in any of the commentaries that the ceiling means something other than its regular meaning (i.e., something lying above). None of the commentaries mention the sky being the ceiling for anything other than the Earth, which is apparent in the statements quoted earlier. It becomes understandable when we read the following verses, **"Or you make the heaven fall upon us in fragments as you have claimed or you bring Allah and the angels before [us]"** [al-Isrāʾ:92]; and, **"And if they were to see a fragment from the sky falling, they would say, '[It is merely] clouds heaped up'"** [al-ṭur:44]; and, **"So cause to fall upon us fragments of the sky, if you should be of the truthful."** Falling off something must mean coming down from a higher position to a lower one.

If we assume the Earth is spherical and roaming in space in this galaxy, where would the sky be? They say it is the space enveloping the Earth, but how would something then fall off the sky on the Earth? Which side of the Earth is facing the sky? It would be all directions, which renders the falling meaningless. They may say that the intended meaning of the verses is subjected to different perspectives. Perish any thought that relativizes the Quranic verses to the extent that it confuses the individual and makes him restless.

At any rate, even if it is tied to one's perspective, all praise is due to Allah that our perspective is consistent with the remaining Quranic verses where Allah states that the sky is a ceiling and that He restrains it from falling on the Earth. The collective number of verses and Arabic words explicitly tell a conscious mind, untainted by heedlessness or the drug of imitation, that the sky is above us like a ceiling. It is neither under the Earth nor envelopes it.

13

QIBLA (PRAYER DIRECTION)

ALLAH, EXALTED AND Glorified, said, "The foolish among the people will say, 'What has turned them away from their qiblah, which they used to face?' Say, 'To Allah belongs the east and the west. He guides whom He wills to a straight path.' And thus We have made you a median [i.e., just] community that you will be witnesses over the people and the Messenger will be a witness over you. And We did not make the qiblah which you used to face except that We might make evident who would follow the Messenger from who would turn back on his heels. And indeed, it is difficult except for those whom Allah has guided. And never would Allah have caused you to lose your faith [i.e., your previous prayers]. Indeed Allah is, to the people, Kind and Merciful. We have certainly seen the turning of your face, [O Muhammad], toward the heaven, and We will surely turn you to a qiblah with which you will be pleased. So turn your face [i.e., yourself] toward al-Masjid al-Ḥarām. And wherever you [believers] are, turn your faces [i.e., yourselves] toward it [in prayer]. Indeed, those who have been given the Scripture [i.e., the Jews and the Christians] well know that it is the truth from their Lord. And Allah is not unaware of what they do. And if you brought to those who were given the Scripture every sign, they would not follow your qiblah. Nor will you be a follower of their qiblah. Nor would they be followers of one another's qiblah. So if you were to follow their desires after what has come to you of knowledge, indeed, you would then be among the wrongdoers. Those to whom We gave the Scripture know him [i.e., Prophet Muhammad] as they know their own sons. But indeed, a party of them conceal the truth while they know [it]. The truth is from your Lord, so never be among the doubters. For each [religious following] is a [prayer] direction toward which it faces. So race to [all that is] good. Wherever you may be, Allah will bring you forth [for judgement] all-together. Indeed, Allah is over all things competent. So from wherever you go out [for prayer, O Muhammad], turn your face toward al-Masjid al-Harām, and indeed, it is the truth from your Lord. And Allah is not unaware of what you do. And from wherever you go out [for prayer], turn your face toward al-Masjid al-Harām. And wherever you [believers] maybe, turn your faces toward it in order that the people will not have any argument against you, except for those of them who commit wrong; so fear them not but fear Me. And [it is] so I may complete My favor upon you and that you may be guided" [Al-Baqarah:142–150].

Muslims have been traveling for more than a thousand years, but they still face a single direction, namely that of al-Kaʿba, regardless of what spot on Earth they are. If the Earth were to be spherical, the directions would not exist. There would not be any north, south, east, or west. Were you to assume that you are at the middle of the spherical Earth on the north side while the Kaʿba is in the middle of the spherical Earth on the south, which direction will the qibla be at? East or west? If you decided to face the qibla from either the north or the south, you will simultaneously be facing and turning your back to the qibla. Do not say you will stand at the nearest spot to you, because I'm saying you are at middle, thus all distances are equal, assuming the presence of the spherical Earth. What will you do?

Dear reader, you should know that if you are on top of the spherical Earth, you will simultaneously be facing and turning your back to the qibla. If they said that a laser ray emits from the individual's head when he prays facing the qibla and reaches the Kaʿba is ridiculous, because if you face either the right or the left direction, the ray will pierce through the Earth. The same is the result if you face south or north unless you maintain that the laser ray bends with the Earth's curvature. The only important issue is that you face al-Masjid al-Harām regardless of the curvatures. The next question is, "Which side do they refer to? Is it the right or the left, the north or the south?" Can you realize the gravity of the dilemma they are experiencing? On the other hand, if the Earth is flat, the direction is always one, and you would not be turning your back to the qibla at any direction and any place on Earth; and all praise is due to Allah.

14

ALLAH SET FIRM MOUNTAINS

IN MORE THAN ten instances in the Quran, Allah, Exalted and Glorified, said that He created firm mountains on the Earth. In the chapter of al-Ra'd, He said, "**And it is He who spread the earth and placed therein firmly set mountains and rivers; and from all of the fruits He made therein two mates; He causes the night to cover the day. Indeed in that are signs for a people who give thought**" [Al-Ra'd:3]. And in the chapter of al-Hijr, He said, "**And the earth—We have spread it and cast therein firmly set mountains and caused to grow therein [something] of every well-balanced thing**" [Al-Hijr:19].

In the chapter of al-Naml, He said, "**Is He [not best] who made the earth a stable ground and placed within it rivers and made for it firmly set mountains and placed between the two seas a barrier? Is there a deity with Allah? [No], but most of them do not know**" [Al-Naml:61]. And, "**And He placed on the Earth firmly set mountains over its surface, and He blessed it and determined therein its [creatures'] sustenance in four days without distinction—for [the information] of those who ask**" [Fuṣṣilat:10]. And, "**And the earth— We spread it out and cast therein firmly set mountains and made grow therein [something] of every beautiful kind**" [Qāf:7]. And, "**And We placed therein lofty, firmly set mountains and have given you to** drink sweet water" [Al-Mursalāt:27]. And, "**And the mountains He set firmly**" [al-Nazi'āt:32].

As you see, there are many verses referring to the firm mountains. You may ask what their purpose is. Allah must have created those mountains for many purposes, most important of which is underscored by Allah in the Quran, as the one in the chapter of Al-Naḥl, "**And He has cast into the Earth firmly set mountains, lest it shift with you, and [made] rivers and roads, that you may be guided**" [Al-Naḥl:15].

What is the meaning of "**shift with you**?" Let us read how exegetes interpreted it.

Al-Baghawy said, "It means to move and sway. The Arabic word '*Mayd* (shift)' expresses imbalance. It is why this is the very word that describes a person with seasickness." In Aḥkām al-Quran by al-Qurtuby, he said, "Mayd is swaying left and right. When the word is collocated with an inanimate object, it means to move; when collocated with tree branches, it means to sway; and when collocated with a person, it means to walk affectedly." Ibn Kathīr commented, "The firm mountains are set to provide stability to Earth and prevent it from shaking with all animals above it so as they lead a stable life. For this reason, Allah said, '**And the mountains He set firmly**.'" Linguists like al-Zaggāg said, "[Mountains are set] to guard against whirling." Ibn Qutayba said, "Mayd expresses a movement and swaying. When collocated to a person's type of walk, it means to slant."

In his commentary, Al-Tabary said, "Allah, Exalted and Glorified, imparted firmness to the Earth through the mountains lest it shakes His creation who inhabit its surface. It was indeed shaking before mountains were set above it. It was narrated to us by Bishr from Yazīd from Saʿīd from Qatāda from al-Hasan from Qays ibn Ubāda who said that the Earth was shaking violently when Allah, Exalted and Glorified, first created it. The angels said, 'This would not provide any stability for anyone above it.' Thereafter it became stable with its own firm mountains. Al-Muthanna narrated to me from al-ḥajjāj ibn al-Minḥāl from ḥammād from ʿAtāʾ ibn al-Sāʾib from Abd Allah ibn ḥabīb from Ali ibn Abu Tālib who said, 'After Allah created the Earth, it was shaky and complained, 'My Lord, why will you place the children of Adam above me to commit major sins and spread impurity above me? Allah then set firm mountains above it, some of which are observable while others are not. The bed of the Earth was shaky." *Mayd* expresses imbalance and swaying. When the word is collocated with a ship, it means the ship is swaying with its passengers. The word *mayd* also refers to the seasickness. Interpreters of the Quran have made similar statements. For example, al-Muthanna narrated from Abu Hudhayfa from Shibl from ibn Abu Najīh from Mujāhid who commented on 'lest it shift with you' as 'it sways with you.' A similar narration was narrated by al-Qāsim from al-ḥusayn from ḥajjāj from ibn Jurayj from Mujāhid."

Al-ḥasan ibn Yaḥya said that Abd al-Razzāq said that Maʿmar narrated to us from Qatāda from al-ḥasan commented on the saying of Allah, "**And He has cast into the Earth firmly set mountains, lest it shift with you, and [made] rivers and roads, that you may be guided,**" by saying, "[Allah] set mountains to prevent the Earth from shaking. Qatāda said that he heard al-ḥasan saying, 'The Earth was shaking after Allah created it. They said, 'This [Earth] would not provide any stability for anyone above it. In the morning, the mountains were created but the angels had no clue from what the mountains were created.'"

Al-Zamakhshary commented, "Mountains were set to prevent the Earth from shaking with you above it. Ma'id refers to the person who feels dizzy from seasickness." Al-Fakhr Al-Rāzī recounted the common statements among exegetes but he objected to it. He said, "According to the popular scholarly opinion on commenting on this verse, a ship sways from one side to the other if it is above water. But when massive objects are placed above this ship, it is stable above water. In the same method, Allah, Almighty, created the Earth above water and it was shaking and shifting but then Allah, Almighty, created the heavy mountains above it to provide it with stability above water by virtue of those mountains."

I say that the scholarly statement is unproblematic in accordance to the perspective of the flatness of the Earth. It is only problematic if we maintain that the Earth is spherical and orbiting or that it is spherical and above water. In his book *Ta'wīlāt Ahl al-Sunna*, Al-Māturīdy said, "[Allah] set the mountains above the Earth lest it shakes with you. Some interpreters said that mountains are set to prevent the shaking because [the Earth] was made flat above water, thus swaying with its inhabitants like a ship above the water. Allah made the Earth firm with the mountains to provide stability for its inhabitants. Critically speaking, however, assuming the Earth was made flat above water, it would not have swayed but sunk into water instead, because the nature of the Earth is to be at the lowest position and sinking under water. On the other hand, if it is maintained that Allah, Exalted and Glorified, created it with an inclination to sway, their opinion would be plausible; and Allah knows best.

"Should they maintain that the Earth was made flat above the wind, it would also be plausible. This reasoning is similar to their statement, 'do you not see that a lantern's light dies inside wells.' It is possible that its light died because of wind at the bottom of the well. We have highlighted this part earlier; and Allah knows best about this. Others have maintained, in addition, that the Earth was made flat above the bull's back and it was swaying by its movements, but Allah made it firm through what He mentioned; and Allah knows best."

As you see, the literal meaning of the Quranic verses states that Allah set the mountains to provide firmness to the Earth and preventing it from shaking and swaying. This would be at odds with the opinion that the Earth is spherical, orbiting around itself and the sun, with a speed up to a thousand miles per hour. Allah, Exalted and Glorified, described the mountains, **"And the mountains as stakes?"** [al-Naba':7]. Stakes function as a means to support the firmness of the tent like the mountains do for the Earth, as the latter provide stability for the Earth so it would not shake.

In Lisān al-'Arab, ibn Manẓūr said, "The stakes of the Earth are the mountains because they make it firm." Obviously, once you fix the tent's stakes, it is only plausible to say that it is firm and not moving. The mountains serve the same purpose for the Earth because they are its stakes. If you were to follow the opinion that the Earth is moving based on the verse, **"And you see the mountains, thinking them rigid, while they will pass as the passing of clouds. [It is] the work of Allah, who perfected all things. Indeed, He is Acquainted with that which you do"** [Al-Naml:88]. Here, Allah had not said that the Earth is the one moving but rather the mountains pass. What is the relationship between the mountains and the Earth, in this regard?

Their perspective is no more than a confusion similar to their opinion regarding the issue of wrapping the night over the day. At any rate, their dedication is incorrect, and the response is explained in the chapter entitled *Refutation of the Allegedly Quran-supported Claim that the Earth is Spherical*. I ask Allah for guidance for me and you.

15

TRAVELING THROUGH THE EARTH

ELIBERATE REFLECTION IN the Quran reveals its invitation to the individual to travel through the Earth to seek admonition. For example, Allah, Almighty, said, "**Similar situations [as yours] have passed on before you, so proceed throughout the earth and observe how was the end of those who denied**" [Āl-'im'rān:137]. He, Exalted and Glorified, said, "**Say, 'Travel through the land; then observe how was the end of the deniers'**" [Al-'An'ām:11]. In the chapter of Yūsuf, Allah said, "**And We sent not before you [as messengers] except men to whom We revealed from among the people of cities. So have they not traveled through the earth and observed how was the end of those before them? And the home of the Hereafter is best for those who fear Allah; then will you not reason?**" [Yūsuf:109]. It is also found in Allah's, Exalted and Glorified, saying, "**So have they not traveled through the earth and have hearts by which to reason and ears by which to hear? For indeed, it is not eyes that are blinded, but blinded are the hearts which are within the breasts**" [Al-Ḥaj:46]. Also, Allah, Almighty, said, "**Say, [O Muhammad], 'Travel through the land and observe how He began creation. Then Allah will produce the final creation. Indeed Allah, over all things, is competent'**" [Al-'Ankabūt:20]. And, "**And We certainly sent into every nation a messenger, [saying], 'Worship Allah and avoid Taghut.' And among them were those whom Allah guided, and among them were** those upon whom error was [deservedly] decreed. So travel through the earth and observe how was the end of the deniers" [Al-Naḥl:36].

Allah, Almighty, also said, "**Say, [O Muhammad], 'Travel through the land and observe how was the end of the criminals'**" [Al-Naml:69]. And, "**Have they not traveled through the earth and observed how was the end of those before them? They were greater than them in power, and they plowed the earth and built it up more than they have built it up, and their messengers came to them with clear evidences. And Allah would not ever have wronged them, but they were wronging themselves**" [Al-Rūm:42]. Allah, Almighty, said, "**Have they not traveled through the land and observed how was the end of those before them? And they were greater than them in power. But Allah is not to be caused failure by anything in the heavens or on the earth. Indeed, He is ever Knowing and Competent**" [Fātir:44]. He, Almighty, said, "**Have they not traveled through the land and observed how was the end of those who were before them? They were greater than them in strength and in impression on the land, but Allah seized them for their sins. And they had not from Allah any protector**" [Ghāfir:21]. And, "**Have they not traveled through the land and observed how was the end of those before them? They were more numerous than themselves and greater in strength and in impression on the land, but they were not availed by what they used to earn**" [Ghāfir:82]. And, "**Have they not traveled through the land and seen how was the end of those before them? Allah destroyed [everything] over them, and for the disbelievers is something comparable**" [Muḥammad:10].

As seen earlier, the command "travel through the Earth" is always present. Throughout the Quran, you would not find "fly through the earth" or any similar statement indicating a human physical movement in the high heavens. Certainly, there are legal reasons for such commands. One, for example, is to take admonition though it is implied. It is as if the Earth is meant for the individual to travel through it whereas he is only allowed to deliberate and think when looking at the heavens or the sky. You can find this in the saying of Allah,

"Say, 'Observe what is in the heavens and earth.' But of no avail will be signs or warners to a people who do not believe" [Yūnus:101]. And, "Who remember Allah while standing or sitting or [lying] on their sides and give thought to the creation of the heavens and the earth, [saying], 'Our Lord, You did not create this aimlessly; exalted are You [above such a thing]; then protect us from the punishment of the Fire'" [Āl-ʿimʾrān:191].

The verses indicate that the human travels through the earth. As for the heavens, it always demands one's deliberation and reflection. In other words, there are no competent creatures like us to take an admonition from since there are no human messages in the heavens nor has anyone been revealed to them therein. The funny part is that some contemporaries cited this verse as a proof of the scientific inimitability of the Quran. They maintained that "travel through the Earth" is an evidence for the existence of the atmosphere, because the human lives on the surface of the Earth but not inside it. The preposition "through" indicates the presence of the atmosphere, as it is a part of the Earth because it moves with it.

I do not know whether one should laugh or cry over such an interpretation. Unfortunately, Sheikh al-Shaʾraawi, whom I dearly love, made a similar state-ment. Of course, some contemporary scholars, like Prof. ʿAdnān Ibrāhim in one of his lectures, have refuted Sheikh al-Shaʾraawi and others. Accordingly, how do you interpret the saying of Allah, Exalted and Glorified, "And the ser-vants of the Most Merciful are those who walk upon the earth easily, and when the ignorant address them [harshly], they say [words of] peace" [Al-Furqān:63]; or His other saying, "Indeed, We have made that which is on the earth adornment for it that We may test them [as to] which of them is best in deed" [Al-Kahf:7]? Also, how would they inter-pret the saying of Allah, Exalted and Glorified, "It is your Lord who drives the ship for you through the sea that you may seek of His bounty" [Al-ʾIsrāʾ:66]? Does the verse indicate above or inside the sea? Furthermore, what would you say regarding the story of Pharaoh's crucifixion of the sorcerers who believed [in Allah], regarding which Allah said, "[Pharaoh] said, 'You believed him before I gave you permission. Indeed, he is your leader who has taught you magic. **So I will surely cut off your hands and your feet on opposite sides, and I will crucify you on the trunks of palm trees, and you will surely know which of us is more severe in [giving] punishment and more enduring"'** [Ṭāhā:71]. He told them he would crucify them on the trunks of the palm trees. Had they been crucified inside the palm trees, above it, or on the air around the trunks of the palm trees? How strange is this interpretation? This interpretation is the outcome of one who attempts to skirt the meaning of Quranic verse to reconcile them with the conclusions of modern science.

A bigot for the flatness of the Earth may argue that the verses inviting to travel through the Earth uses the preposition through instead of on to indicate that the Earth has high edges and the inhabited parts are lower than the edges, which proves that the creatures above it are traveling through it, because they cannot get outside of it given the high edges created by the flatness of the Earth. Undoubtedly, this is significantly more logical than the view proposed by those influenced by the conclusions of modern science. However, it should be maintained that traveling is one thing and walking is another thing.

On this regard, I would like to quote an Arabic-related rhetorical point highlighted by Prof. Fāḍil al-Sāmirrā'y (may Allah bless him). He said, "Linguistically speaking, walking is different from traveling. Walking is used to describe a movement from a certain location to a specific destination. In the Quran, walking has a purpose either for seeking admonition or for a trade or any other reason, as in the saying of Allah, Almighty, '**And when Moses had completed the term and was walking with his family**' [Al-Qaṣaṣ:29]. Walking here is from Madyan to Egypt; it is not literally walking. Other meanings of walking in the Quran includes walking for a long distance to seek admonition, as in the saying of Allah, Almighty, '**Say, 'Travel through the land; then observe.**'' This meaning of walking is mentioned in eleven verses in the Quran as in '**So have they not traveled through the earth**;' and, '**Say, [O Muhammad], 'Travel through the land and observe.**'' Traveling here means move across land. Once again, it is mentioned in the Quran to indicate walking for a long distance for the purpose of trade or seeking admonition. Mere walking, however, expresses the bare movement, but not necessarily heading to a particular goal, as in Allah's, Almighty, saying, '**And the servants of the Most Merciful are those who walk upon the earth easily.**'"

How many times have you discussed with the modern-day imitators about any issue, whether Sharia-related or scientific, to which the response was to demand you to ask specialists and experts without philosophizing? The problem of modern-day imitators of the contemporary scientific society is that they have nicknamed some of the atheists and arrogant people with Quranic names only because they have been prominent on particular issues and considered them from a single angel. They have not labored to deliberately investigate the Quran to discover the truth of the matter. For example, they classify Stephen Hawking or Sean Carrol, including the less prominent ones like Neil deGrasse Tyson and Richard Dawkins, as sage scholars and insightful people. We are familiar with their points of view about Allah and Islam and how derisive and inferior they consider them. I invite anyone who is unfamiliar with them to read and follow their work, because they are available today. Are these people really the sages?

Let us read what the Quran says about the true sages, "**Indeed, in the creation of the heavens and the earth and the alternation of the night and the day are signs for those of understanding**" [Āl-'im'rān:190]. The Quran highlights those with understanding, not atheists. The former's identification in the Quran is, "**Who remember Allah while standing or sitting or [lying] on their sides and give thought to the creation of the heavens and the earth, [saying], 'Our Lord, You did not create this aimlessly; exalted are You [above such a thing]; then protect us from the punishment of the Fire**'" [Āl-'im'rān:191]. These are the conditions that must be present in those who have minds. Now, does any of those descriptions apply to any of the scientists I mentioned earlier? Or do we find arrogance and other blameworthy characteristics present in them? Consider what Allah, Almighty, said about the arrogant people. "**I will turn away from My signs those who are arrogant upon the earth without right; and if they should see every sign, they will not believe in it. And if they see the way of consciousness, they will not adopt it as a way; but if they see the way of error, they will adopt it as a way. That is because they have denied Our signs and they were heedless of them**" [Al-'A'rāf:146].

If Allah turns them away from His signs, how could they guide you to the signs of Allah on Earth and on the sky? Does this not mean that the disbelieving atheists are heedless and unaware of the signs of Allah? So, why would you be happy with them? Is it because those currently so-called scientists, who knew the truth but denied it, are exempted from such action of turning away from the signs of Allah? Consider carefully the verse I quoted, **"I will turn away from My signs**." It is as if the signs are present, but they were turned away from perceiving it. It does not mean the signs came while they were attempting to perceive them. The signs are ever present in their respective locations.

16

QUICK REFLECTIONS ON THE QURAN

THE FOLLOWING PAGES shed light on some quick reflections on some Quranic verses in a bid to inspire you to comprehend myriad matters in light of these salient divine signposts.

Allah, Almighty, said, **"If you asked them, 'Who created the heavens and earth and subjected the sun and the moon?' they would surely say, 'Allah.' Then how are they deluded?"** [Al-'Ankabūt:61].

The abovementioned verse indicates that the people at the time of the Messenger (Peace and blessings be upon him) were familiar with the creator of the heavens, and they knew the meaning of the heavens. Otherwise, they would have decried it and found it very odd. Hence, such was the case back then. Dear readers, I would like to you: Did they have at the time the state-of-the-art technology that could have enabled them to discover what is beyond the sun, the moon and the known seven planets or to see what is beyond the galaxies? Or was their vision limited? Since their vision was limited, how come they knew the heavens unless they were very aware of the shy and they knew it very well.

The following verse further substantiate that. Allah, Almighty, said, **"[More precisely], is He [not best] who created the heavens and the earth and sent down for you rain from the sky, causing to grow thereby gardens of joyful beauty which you could not [otherwise] have grown the trees thereof? Is there a deity with Allah? [No], but they are a people who ascribe equals [to Him]"** [Al-Naml:60]. In actuality, the heavens were human beings to the people back then, and this is what the Quran has repeatedly reiterated. Allah, Exalted and Glorified, said, **"And if you asked them, 'Who created the heavens and earth?' they would surely say, 'Allah.' Say, '[All] praise is [due] to Allah;' but most of them do not know"** [Luqmān:25]. Allah, Almighty, also said, **"And if you asked them, 'Who created the heavens and the earth?' they would surely say, 'Allah.' Say, 'Then have you considered what you invoke besides Allah? If Allah intended me harm, are they removers of His harm; or if He intended me mercy, are they with-holders of His mercy?' Say, 'Sufficient for me is Allah; upon Him [alone] rely the [wise] reliers'"** [Al-Zumar:38]. Allah, Almighty, further said, **"And if you should ask them, 'Who has created the heavens and the earth?' they would surely say, 'They were created by the Exalted in Might, the Knowing'"** [Al-Zukhruf:9]. He, Exalted and Glorified, also said, "And He has subjected to you whatever is in the heavens and whatever is on the earth—all from Him. Indeed, in that are signs for a people who give thought" [Al-Jāthiya:13]. The people know these created celestial bodies, and that is why it is said that these creations are favors bestowed by Allah upon them, are they not?

Allah, Exalted and Glorified, said, **"And if whatever trees upon the earth were pens and the sea [was ink], replenished thereafter by seven [more] seas, the words of Allah would not be exhausted. Indeed, Allah is Exalted in Might and Wise"** [Luqmān:27]. Leaving aside the linguistic inferences, let's ask ourselves. Why did He say, "the sea replenished thereafter" instead of saying "the sea replenished from more seas around it?" How can it be beyond it from the point of view of those who claim that the earth is spherical? It is impossible to imagine that. However, if we consider the earth flat and that there is a sea surrounding it, it is not impossible that this sea is replenished thereafter by seven more seas. Interestingly, when such a supposition came across my mind, I thought I was the only one who thought that there is a sea surrounding the inhabited earth. Yet, as usual, I found a number of the very early generation of Muslims had reached the same conclusions well before me.

For instance, we find in the commentary of Ibn Kathīr (May Allah have mercy on him) a recorded narration attributed to Ibn ʿAbbās but with a disconnected chain of transmission. It reads, "On the authority of Ibn ʿAbbās who said, 'Allah created beyond this earth an oceanic seas surrounding it, and He created beyond that a mountain that is seven times the size of this earth. Then, He created beyond all of that a sea surrounding it. Then, He created beyond all of that a mountain dubbed 'Qāf' where the second heaven stands as a roof over it.' He went on enumerating seven earths, seven seas, seven mountains, and seven heavens. Then he commented that this is the meaning of Allah's statement, 'the sea [was ink], replenished thereafter by seven [more] seas.'"

Ibn Kathīr stated elsewhere in his commentary in response to the idea of the existence of the seas, "There are no seven seas surrounding the earth as claimed by those who knew about that from the Israelite tales that can neither be held true nor false. The truth is what Allah, Almighty, related in the other verse, **'Say, 'If the sea were ink for [writing] the words of my Lord, the sea would be exhausted before the words of my Lord were exhausted, even if We brought the like of it as a supplement''** [Al-Kahf:109]. His statement 'the like of it' does not mean just one more sea but rather another sea followed by another one ad infinitum because Allah's signs and words are innumerable."

Hence, Ibn Kathīr's statement is a clear indication that Quran commentaries are fraught with untold Israelite tales. Such Israelite tales are not easily verified. Moreover, the recorded narration was attributed to IbnʿAbbās but with a disconnected chain of transmission that could serve as a lens to conjure up as of this universe. That does not necessarily mean this is the conception of the universe I have faith in. However, it is a conception that is worth further examination and research.

If we accept that the heavens are seven and the earths are seven as well, and if we accept that our earth is the smallest earth that is situated in the middle of the cosmos, that it is the lowest and the deepest earth, that there is an oceanic sea that surrounds it, and that a mountain dubbed Qāf encircles that oceanic sea, and if we accept that the seven heavens are like a dome on earth, then the edges of the dome of the first heaven converges with the sides of the Qāf mountain. The second earth lies after that mountain and there is an oceanic sea surrounding it. The oceanic sea in turn is surrounded by the Qāf mountain that encircles it. The dome of the second heaven stands as a dome on this earth and its edges come together with the Qāf mountain that encircles it. The same applies to the rest of the seven heavens and seven earths. If the seven earths are arranged in layers with each one constituting a layer on its own, then the heavens have seven layers that correspond to the seven layers of the earth. As such, one can reconcile this conceptualization with the Quranic verses, the prophetic hadiths, and the myriad narrations attributed to the Companions. However, is this the one conceptualization that these three (i.e., the Quranic verses, the hadiths, and the recorded narrations) propound? Of course, no. There are a number of other hypotheses that I have set forth in this volume.

Allah, Almighty, said, "**Indeed, we offered the Trust to the heavens and the earth and the mountains, and they declined to bear it and feared it; but man [undertook to] bear it. Indeed, he was unjust and ignorant**" [Al-Aḥzāb:72].

The verse suggests that the material creations are the heavens, the earth, and the mountains. Notice that He did not mention the seas and the like as if they are special creations. If it is argued, "Why did He not mention the sea?" In response, I say that the sea is not a solid object. If they object, "What about the heavens?" I say that He already responded to that question by declaring that the sky is actually a ceiling. On the whole, I have mentioned in an earlier chapter that the Earth includes both the dry land and the water. It may not encompass both in some verses, depending on the context, the backdrop, and so on. You can review that in the corresponding chapter.

Moreover, one should think anew about these three things in the statement of Allah, Exalted and Glorified, "**Then do they not look at the camels—how they are created? And at the sky—how it is raised? And at the mountains—how they are erected? And at the earth—how it is spread out?**" [Al-Ghāshiya:17–20]. I shall elaborate more on "the camels" later. Let us now reflect on His saying, "**And at the sky—how it is raised?**" [Al-Ghāshiya:18]. Is not this the sky that stands high above us? If so, why did He say it was raised high? Did he mention that because it was raised in a certain way? Is there a particular rationale behind mentioning that He raised it? What is customary for human beings is that what was raised high above stands on something that supports it and prevent it from falling down, whether this raised thing is tangible or intangible. It has to have a base and a foundation as well as a support that raises it. The sky and the Earth are similar to a house or a building that has a foundation and a ceiling. The ceiling is the sky. However, the sky is raised high without pillars that we could see.

On another level, there is one more issue here. We cannot say that something is raised unless it is raised high above something else that is under it. This is impossible to comprehend if we accept that the Earth is spherical, that it is a tiny dot that moves around in space, and that the sky surrounds it from all directions. The sky would then enfold from below the southern part of earth but, by virtue of directions, it seems raised high above for those living on that southern part. Yet, this relative hypothesis is not the truth, per se. In theory and praxis, this vision better fits the conceptualization of the Earth as flat and, in fact, it exquisitely conforms to it because the Earth has no sky under it. The sky stands high above; namely, it is raised. That is why, it is more apt to wonder how it was raised as such. As it stands, it is even better to reflect on the signs it has to offer. After all, Allah knows best.

As for His saying, **"And at the earth—how it is spread out?"** [Al-Ghāshiya:20], its meaning is crystal clear. It is true that this verse could be so from the perspective of human beings. Yet, in fact, humans sometimes see mountains, plateaus, and so forth on Earth, so how can it be said that it is flat? The answer is that if you were to look at the Earth in its entirety and see it from above it, from beyond it, or from outside of it, you will then know that it is flat, in spite of the existence of mountains, hills and so on.

As for His saying, **"Then do they not look at the camels"** [Al-Ghāshiya:17], commentators provided numerous interpretations regarding the reasons for mentioning the camels in this context, and definitely some of these interpretations hold true, God willing. The bulk of these are remarkable, lucid, and comprehensible to the average reader. However, I have an additional remark that I still do not know whether they are accurate or not, but still I want to present them here to you.

It seems to me that the shape and form of the Arabian camels somehow resemble the quintessential formation of the universe in which humans live, namely the heavens and the Earth. The camels' hump looks like a mountain and like the dome of the sky resembles a tent set up on earth. Likewise, the four legs are the four pillars of earth and its stakes. The abdomen of the camel is the Earth, and it stores supplies just as the Earth does. The bizarre neck of the camel epitomizes the creature that surrounds that Earth. It is also easy to move around, just as it is easy to cause the Earth and all that is on it to revolve around. It is possible to put a saddle on the back of the camels just as the houses established on Earth or the throne on the heavens. Furthermore, the creation of the camel could also juxtapose with that of the Earth, but this requires a lengthy explanation. As for the creature that surrounds that universe, talking about it raises many eyebrows and leaves many mouths agape. I will defer discussing what I have in mind about it later on given that this volume is meant to discuss a completely different matter. This book is aimed at readers curious to know about the shape of the Earth.

Allah, Exalted and Glorified, said **"Have We not made the earth a container"** [Al-Mursalāt:25].

Have you ever wondered what the term *kifātan* (a container) means? The meaning of the verse is crystal clear, but there is no harm in reflecting on it. What does *kifātan* mean? Literally, as Ibn Jarīr al-Ṭabarī argues, "the term *kifātan* means a container as one would say: This is a *kift* or *kifīt* for that to refer to its container." Al-Zamakhsharī maintained, "the term *al-kifāt* means to amass or collect things together. This is similar to the form and purport of such terms as *al-ḍimām* and *al-jimā'*. They say: This chapter is the *al-jimā'* (summary) of all chapters." Ibn 'Āshūr stated, "the term *al-kifāt* refers to the thing in which things are amassed and collected. It is a noun derived from the verb *kafata,* which means to gather and collect, and whence the container is dubbed *kifātan* just as *al-wi'ā'* (a container) is dubbed as such because it holds things inside it and *al-ḍimām* (a holder) for the things that hold collected things inside it."

As such, we deduce that the Earth is a container for the dead beings. But how can the spherical Earth be a container for the living beings? If you understand that the Earth is flat and its edges are higher than the creations on it, then this explains how the Earth is like a container that holds us, gathers us together, and prevent us from going out from it. This verse is a clear indication of that, as it was the case with the verse that I elaborated on earlier. Moreover, this meaning exquisitely conforms to the saying of Allah, Exalted and Glorified, "**O company of jinn and mankind, if you are able to pass beyond the regions of the heavens and the earth, then pass. You will not pass except by authority [from Allah]**" [Al-Raḥmān:33].

Allah, Almighty, said, "**And at the earth—how it is spread out?**" [Al-Ghāshiya:20]. He described it as flat and not spherical. Ibn Jarīr al-Ṭabarī maintained, "He says: And at the earth—how it is levelled out." People literally describe a mountain as *muṣaṭah* if it is flat at the top. The scholars of exegesis subscribed to this view of ours. As far as those who espoused this view are concerned, Bishr narrated that Saʿīd reported on the authority of Qatāda, who commented on the verse "**And at the earth—how it is spread out?**" saying that it means "leveled out." It seems as if Allah is saying, "Is not the One who created this capable of creating what He pleases in paradise?" Ironically, those who opine that the Earth is spherical attempted to make this verse consistent with their view, as Al-Rāzī did in his well-known interpretation that his successors unthinkingly espoused.

In this regard, I would like to ask those who would land on the moon and would occupy Mars, those who see the Earth as a dot flying non-stop in space, how do they feel when they read the following verses in the Quran, "**And at the earth—how it is spread out?**" [Al-Ghāshiya:20]. "**And the earth— We spread it out**" [Qāf:7]. "**And when the earth has been extended**" [Al-Inshiqāq:3]. "**[He] who made for you the earth a bed [spread out]**" [Al-Baqara:22]. "**Have We not made the earth a resting place?**" [Al-Nabaʾ:6]. These are but a few verses that indicate the Earth is flattened out. How do they feel when they see, everywhere they go, the Earth on television and the globe miniatures in shopping malls and even in their own offices where they work? What kinds of feelings do their hearts undergo? Would they not argue with you, Muslims, that they see the Earth as a spherical ball in all images? Will the following verses apply to those who claim they will live in space? "**Then do they not look at the camels—how they are created?**" to "**And at the earth—how it is spread out?**" [Al-Ghāshiya:17–20]. The verb "look" is in the present case and denotes continuity, implying this will truly be the case till the end of times. How will they comprehend these verses if the Earth is not flat before them? Cannot they simply argue that they do not see it flattened out before them and instead claim that they see it as an insignificant ball that flies in space?

If astronauts state they saw it as a spherical ball and they take fabricated photos of it and if lay people on Earth say that they see it flattened out for them, what will that bring about in the hearts of future generations? But for Allah's guidance and mercy, we will indeed be in trouble.

Allah, Almighty, said, "**[The one] who has made for you the earth a bed and made for you upon it roads that you might be guided**" [Al-Zukhruf:10]. The term *mahdan* (a bed) explicitly and implicitly means that the earth is flat and not spherical. He, Almighty, also said, "**And the earth— We spread it out and cast therein firmly set mountains and made grow therein [something] of every beautiful kind**" [Qāf:7]. He said *madadnāhā* (flattened it out) and not made it spherical. He, Almighty, also said, "**And the earth We have spread out, and excellent is the preparer**" [Al-Zāriyāt:48]. He said He spread it out, not made it spherical. He, Almighty, also said, "**Have We not made the earth a resting place?**" [Al-Nabaʾ:6]. He stated that He spread it out, flattened it out, and leveled it out like a bed. Could any of these mean it is spherical?

Allah, Almighty, said, "**And [by] the earth and He who spread it**" [Al-Shams:6]. Ibn Jarīr al-Ṭabarī maintained, "This verse too is contrasted with the earlier one. The verse means that by the Earth and the One Who spread it out. The term *ṭaḥāha* means flattened it out on the left, on the right, and from all directions. Scholars of exegesis differed regarding the meaning of *ṭaḥāha*, with some commentators arguing it means, "By the earth and what He created on it." As far as those who espoused this view are concerned, Muḥammad ibn Saʿd narrated from his father who narrated from his uncle who narrated that his father narrated from his father on the authority of Ibn ʿAbbās who said that this above-mentioned verse [Al-Shams:6] means that by the earth and what He created on it.

Other commentators argued that it means that by the Earth and how He flattened it out. The following are the ones who opined that. Muḥammad ibn Umāra narrated to us on the authority of ʿUbayd Allāh ibn Mūsa, who reported from ʿIssa who narrated to him from al-Ḥārith from al-Ḥassan who related from Warqāʾ who all narrated it on the authority of Ibn Abū Najīḥ from Mujāhid, said regarding **"And [by] the earth and He who spread it,"** i.e., flattened it out. Yunus narrated that Ibn Wahb related that Ibn Zayd said regarding, **"And [by] the earth and He who spread it"** that it means levelled it out.

Other commentators offered a different line of argument, contending that the verse refers to "the Earth and how He divided it." Among those who espoused this view are ʿAlī who narrated on the authority of Abū Ṣālih who narrated from Muʿawiyya from ʿAlī who narrated that Ibn ʿAbbās said that this abovementioned verse [Al-Shams:6] means "by the Earth and how He partitioned it."

On another level, al-Qurṭubī remarked, "The term ṭaḥāha means flattening it out or the One Who flattened it out, as we have elaborated earlier, meaning levelling the earth out." Ibn Kathīr also said it means, "He spread it out, and this is the most preferable opinion and the majority of commentators have espoused this view. Linguists have subscribed to this very same view as well." Al-Jawharī maintained that the term ṭaḥāha is synonymous with the term daḥāha which means "to spread something out." Al-Fakhr al-Rāzī said, "This verse has two points to consider. First, He mentioned this verse after His saying, **'And [by] the sky and He who constructed it'** [Al-Shams:5], because of His other saying, **'And after that He spread the earth'** [Al-Nāziʿāt:30]. Second, al-Layth maintained that the term ṭaḥāha is like the term daḥāha in the sense that they mean spreading something out. Lingually, substituting the letter 'ṭ' with the letter 'd' is also plausible, and it means 'He extended it.' ʿAtāʾ and al-Kalbī said, 'He spread it out on the water.'" In his commentary Aḍwāʾ al-Bayān, al-Shinqīṭī stated that, "there is no contradiction in that He, Almighty, created it, extended it, and spread out its edges in all directions."

Allah, Almighty, said, **"And [mention, O Muhammad], when your Lord said to the angels, 'Indeed, I will make upon the earth a successive authority.' They said, 'Will You place upon it one who causes corruption therein and sheds blood, while we declare Your praise**

and sanctify You?' Allah said, 'Indeed, I know that which you do not know'" [Al-Baqara:30].

This verse clearly indicates man was Allah's vicegerent on earth. That is to say human beings do not have to be vicegerents on other planets in space. I really have no clue what humans are trying to achieve from vying to conquer space and occupy planets. Allah, Exalted and Glorified, did not say that He is making man His vicegerent in the heavens and on Earth, but, rather, He said, **"I will make upon the earth a successive authority."** Therefore, relentless human attempts to live in space are some sort of childish ambitions and wishful thinking, to say the least. Human beings should know they are meant to be Allah's vicegerents here on earth, and not on Mars or on any other planet. As for those who argue that the Earth is a term that could be applied to any other celestial body other like this one on which we live to any planet in the sky or the like, their argument carries no weight whatsoever because theirs is an absurd argument that is indicative of not only their limited understanding but also of their very mediocre comprehension that makes me refrain from even engaging in refuting such an unadulterated nonsense. Luckily, commentators with linguistic backgrounds have already offered rebuttals for that, thus relieving me from undertaking such an endeavor.

As a matter of fact, this verse made me realize the paramount salience of the Earth, given the response of the angels that suggests the Earth received special attention. I am familiar with all that has been said or written regarding the reply of the angels. Yet, I still have not read about the possibility that the angels might have wanted to be Allah's vicegerent on Earth, that is, to live on Earth? It is possible, is not it? Maybe! Let us assume that, especially if we can assume the question that should be raised in this respect: Why are the angels very interested in the Earth if it were to be a tiny dot in space that floats in the galaxy? Given that they live in the higher heavens, what makes them so interested in it? Does that not show how important the Earth is, even for the angels? I cannot imagine that such a salient position is accorded to the insignificantly tiny spherical earth compared to the sky. However, if we were to argue that the size of the Earth is equal to that of the heavens that stand high above it, things become crystal clear, and it becomes conceivable that all creations yearn to serve as vicegerents on it.

Allah, Almighty, said, "He said, 'O Adam, inform them of their names.' And when he had informed them of their names, He said, 'Did I not tell you that I know the unseen [aspects] of the heavens and the earth? And I know what you reveal and what you have concealed'" [Al-Baqara:33].

This verse reveals that there are things on Earth that are still unknown. This is a counterargument to those who claim that modern man and the scientific community have known everything on Earth. To know the ins and outs of everything on Earth is impossible. Let us look at the deepest thing human beings have known in the Earth or the sea! The deepest hole ever known and recorded inside the earth was the Kola Superdeep Borehole in Russia, which is 12 km deep. What a trivial depth! This is the deepest that human beings as well as their cutting-edge equipment and advanced technology have to offer. As for the deepest point in the sea, it is the well-known Project Nekton where they reached around 10 km only below the bottom of the sea. Meanwhile, you see some people are developing theories on countless unbounded numbers. For instance, they divided the earth into three layers, and they split them all into seven. The first layer is the solid crust. Compared to other layers below it, it is like the outer crust of an orange outside the ones we eat. Just imagine that the deepest they we explored below the surface of the Earth or the bottom of the sea is only 13 kilometers. Then, if you were to determine the distance between us and the moon or the sun, it will be a task of gargantuan proportions. Alas, this is what they teach in schools. Ask them, "Why is not there a record of anyone who walked around the Earth from the north to the south or vice versa?" They have no answer to this question. Such flimsy answers as inclement weather conditions, severe temperatures, and the inexistence of technology come to no avail. Such shaky statements should be dumped into the closest dustbin since they themselves claimed that they had landed on the moon more than fifty years ago.

When human beings failed to know that which is close to them, they turned the attention of their fellow humans to such unadulterated nonsense. Alas, some people who were granted guidance from Allah and a glorious scripture failed to ask themselves the question that should be posed to the so-called luminaries of the scientific community: "Since you have not discovered the unseen secrets of the earth, why are you so preoccupied with exploring the sky?" Dear readers, have you ever wondered whether or not the countless unbounded numbers reiterated in scientific periodicals are premised on scientific theories or mere hypotheses? Do you know, dear readers, what the difference is between a theory, a hypothesis, and a fast and fixed rule, as well as other frequently used scientific jargon? If you were to realize that the bulk of us are not familiar with the difference between a hypothesis and a theory from a scientific point of view, let alone bother themselves to look into that in the first place, you should ask yourselves, "Why do they all believe the Earth is spherical, and why do they consider that as an indisputable fact? Did they reach such a conclusion based on in-depth research and inquiry? Or is it a manifestation of the blind imitation and the mindless belief in what the West has come up with?" As a matter of fact, try to ask yourself as well as those around you such salient questions and you will come to realize the distressing situation we are stuck in.

Allah, Almighty, said, "**But Satan caused them to slip out of it and removed them from that [condition] in which they had been. And We said, 'Go down, [all of you], as enemies to one another, and you will have upon the earth a place of settlement and provision for a time'**" [Al-Baqara:36].

We understand from this verse that the settling place for human beings is the Earth, and not Mars nor any other planet. He did not say that there is an abode for you on Earth but rather said that "**you will have upon the earth a place of settlement,**" which is a word that suggests continuity and uninterrupted existence. Likewise, the verse indicates that means of sustenance will not be available on other planets, regardless of what they call them. By the same token, He reiterated the same meaning elsewhere in the chapter of Al-A'rāf, where He said, "**[Allah] said, 'Descend, being to one another enemies. And for you on the earth is a place of settlement and enjoyment for a time'**" [Al-A'rāf:24]. He said "**a place of settlement**" and not just a place where you can stay for a temporary period of time and then move elsewhere. Perhaps, one day, they will conquer space and build communities and colonies in space. Furthermore, the same applies to His saying, "**And it is He who** has multiplied you throughout the earth, and to Him you will be gathered" [Al-Muʾminūn:79]. This verse refers to the story of human beings on Earth. Likewise, His saying, "[Allah] will say, 'How long did you remain on earth in number of years?'" [Al-Muʾminūn:79], refers to what takes place on the Day of Judgment. If human beings were to live elsewhere on other planets in the sky, He would have mentioned that too. There is no linguistic indication that the Earth refers to the celestial bodies by large. Thus, the unfounded claims of those weak-kneed people were utterly wiped out. Allah, Exalted and Glorified, said, "It is He who made you [people] successors to the land. Those who deny the truth will bear the consequences: their denial will only make them more odious to their Lord and add only to their loss" [Fāṭir:39]. He made you vicegerents on this earth and not somewhere else. Allah, Exalted and Glorified, further said, "Say, 'It is He who has multiplied you throughout the earth, and to Him you will be gathered'" [Al-Mulk:24]. He dispersed you all over the Earth and not over somewhere else.

Allah, Exalted and Glorified, said, "Originator of the heavens and the earth. When He decrees a matter, He only says to it, 'Be,' and it is" [Al-Baqara:117]. This verse proffers a clear indication that Allah created the Earth like no other. This verse is a rebuttal to the arguments of those who were influenced by the falsified arguments that this universe came into being by accident or that this universe came about after millions of millions of trials and errors, and to those who were influenced by Hindu notions that million similar versions of this universe had existed before. If that was the case, Allah, Almighty, would not have said, "Originator of the heavens and the earth." Perhaps this is a nuanced point to reflect on, particularly for the minds of researchers who are interested in such nuances and fun facts.

Allah, Almighty, said, "Indeed, in the creation of the heavens and earth, and the alternation of the night and the day, and the [great] ships which sail through the sea with that which benefits people, and what Allah has sent down from the heavens of rain, giving life thereby to the earth after its lifelessness and dispersing therein every [kind of] moving creature, and [His] directing of the winds and the

clouds controlled between the heaven and the earth are signs for people who use reason" [Al-Baqara:164].

As a matter of fact, I cannot deny that I have reflected a lot on this verse, because my mind took me very far. There is no harm to share with you, dear readers, some of the ideas that came across my mind while I was reflecting on this aforementioned verse. In this verse, Allah, the Exalted and Glorified, enumerates His signs to the people including the creation of the heavens and the Earth, the alternation of the day and night, and so forth, as spelled out in the verse. Most importantly, what made me pause to reflect on this verse is the mention of the ships that sail through the sea. After the mention of the ships that sail through the sea, He mentioned the rain and then the shifting of the winds. That made me wonder, "Why does Allah mention the ships in the midst of enumerating His upper and lower signs and favors (the heavens, the day and night, the winds, and the clouds), and why does He not mention the animals, the fauna, and the like?" Thereupon, I entertained the thought that the term *falak* in this verse does not mean the ships but rather the spherical celestial bodies that are in the sky, that the sky is indeed a sea, and that these celestial bodies float in it. There are sundry verses in which the stars were depicted as signposts and marks that guide people on Earth and in the sea. The same applies to the sun and the moon. I felt astounded when such a thought came across my mind and felt moved by it, despite thinking it was such a very far-fetched opinion that is far from right.

Then, I carried out a comparison between the *falak* (ships or celestial bodies in space) and the sun and the moon. I read the following verses, "**And it is He who created the night and the day and the sun and the moon; all [heavenly bodies] in an orbit are swimming**" [Al-Anbiyā':33], and "**It is not allowable for the sun to reach the moon, nor does the night overtake the day, but each, in an orbit, is swimming**" [Yāsīn:40]. Then, I asked myself, "What is the relationship of the sun and the moon with the *falak* (ships)? Why do they float as if they are swimming in a sea?" I am also familiar with the numerous meanings of the term *tasbīḥ* (*lit.* to exalt and glorify Allah) as argued by the commentators with all their schools of thought, orientations, iterations, and ideologies. Therefore, you should not stop here thinking that I am not acquainted with these interpretations, and you should continue reading.

Then, I came across this verse that made me pause to reflect on it: "**And the sun runs [on course] toward its stopping point. That is the determination of the Exalted in Might, the Knowing**" [Yāsīn:38]. Thereupon, I exclaimed, saying, "Glory be to Allah!" There, He said, "**and al-fulk which sail through the sea,**" and here, He said, "And the sun runs [on course] toward its stopping point." My mind went very far, reflecting on these verses, and I thought about ideas and theories most readers know little or nothing about. I will keep these reflections to myself, for now, on the hope that I will jot them down someday in the future. Expounding on these ideas requires a lengthy discussion and a great deal of concentration on the part of readers. This compact volume on the issue in question is not meant for that topic. The reason I wrote this is that we ought to pause and reflect on this Quranic verse, study it very well, and try to fully comprehend it. If there are attempts made, then the outcomes will be mostly positive. What I have mentioned does not necessarily mean the term *falak* means what I have suggested. The term *falak* in numerous verses refers to the ships that people use to sail through the sea, and Allah knows best. As for the issue of whether or not there is a sea in the sky, I shall elaborate on it in detail later on.

Allah, Almighty, said, "**Allah—there is no deity except Him, the Ever-Living, the Sustainer of [all] existence. Neither drowsiness overtakes Him nor sleep. To Him belongs whatever is in the heavens and whatever is on the earth. Who is it that can intercede with Him except by His permission? He knows what is [presently] before them and what will be after them, and they encompass not a thing of His knowledge except for what He wills. His *Kursi* extends over the heavens and the earth, and their preservation tires Him not. And He is the Most High, the Most Great**" [Al-Baqara:255].

We often notice that Allah, Exalted and Glorified, mentions the Earth in contrast with the sky. Today, the scientific circles keep relating to us tales of the gargantuan size of the universe and that the Earth is no more than a tiny speck of sand in this extremely vast universe. Even more, the scientific community claims that the already-discovered universe of ours today encompasses around two hundred billion to two trillion galaxies! They further claim that the average number of stars in each galaxy is one hundred stars, some of which is bigger than the sun while others are smaller than it. And what will make you realize how big the sun is for them? They claim that the size of the sun is one million, three hundred thousand times the size of the Earth! They also argue today that there are only seventeen billion Earth-sized stars in our galaxy, which is dubbed the Milky Way.

As I have explained earlier, they go as far as saying that there are 200 billion galaxies in our universe, so do the math! What is the Earth like in such extremely vast universe? Definitely, no more than a tiny atom. This is what modern science promulgates. On the contrary, we see that the Earth in the Quran is depicted as very significant, not only in terms of position but in terms of size as well. As we have seen in the aforementioned verse, the throne extends as wide as the heavens and the Earth. Scholars of the Arabic language know that the rule is that a significant thing should not be exposed with an insignificant thing. Exceptions are very far and few between. As a rule of thumb, we can hardly find a verse in the Quran that mentions the heavens without accompanying it with the Earth. Why is that the case? The reason is that the Earth is not a star or a planet. It is a creation other than the planets, the stars, the sun, and the moon. It is a completely different creation that it is like no other. It seems to me, based on my reflections and inquiry, that the heavens and the earth constitute what the people call today the universe. Put differently, the heavens and the Earth encompass the unfolding narrative of humanity and its objectives. If this were not the case, then where can we find in the Quran that the Earth was described as a planet, a start, or the like? Not a single verse in the Quran attests to such confusion that the scientific community and their mindless Muslim and Arab henchmen make. Since there is nothing in the Quran that indicates the Earth is a planet or a star, how did Muslims conclude that the Earth is a planet or a star?

Kindly bear with me. They argue that since they see the sun, the moon, and some planets as spherical, then the Earth too should be spherical! Oh, my goodness! Dear readers, we are clearly going through a real mental and intellectual plight. We really do, otherwise we would not have premised our conclusion that the Earth is spherical on such pitiful, pathetic, and laughable deduction. Indeed, there is no power and might except of Allah's Who is the Best Helper.

Kindly read what Allah, Exalted and Glorified, said with regard to the planets and stars, **"And We have certainly beautified the nearest heaven with stars and have made [from] them what is thrown at the devils and have prepared for them the punishment of the Blaze"** [Al-Mulk:5]. Is the Earth like that? Is it a lamp-like star? Do they not say that the Earth is dark, and it is the sun that gives rise to the alternation of the day and night? Has the Earth ever been made as a missile for stoning eavesdropping devils? Dear readers, have you ever read or heard that the Earth stones eavesdropping devils or shoots a burning flame? Even more, how can the Earth stone eavesdropping devils when it in and of itself is an abode for many devils from both human beings and jinn? Allah, Exalted and Glorified, said in the chapter of Al-Ṣaffāt, **"Indeed, We have adorned the nearest heaven with an adornment of stars. And as a protection against every rebellious devil"** [Al-Ṣaffāt:6–7]. The interpretation of this verse is similar to the above-mentioned one. Therefore, the claim that the Earth is a planet is far from right. If the Earth were to be one of the stars or the planets that Allah created to decorate the sky with, then why do we find myriad verses in the Quran that point out the exposition of the sky versus the Earth, or of the heavens versus the Earth, as well as what is in between them? Even hadiths inform us about the distance between the sky and the Earth. This is a common sense for those who read a few Prophetic traditions.

Allah, Exalted and Glorified, said, **"Indeed, those who disbelieve and die while they are disbelievers—never would the [whole] capacity of the earth in gold be accepted from one of them if he would [seek to] ransom himself with it. For those there will be a painful punishment, and they will have no helpers"** [Āl-ʿImrān:91].

As a matter of fact, when I read this verse while I was examining the shape of the Earth, I imagined the shape of the Earth as a container, or at least as something that has raised sides. For instance, one would say: a container filled with water or the cup is full and so on. But does that mean the Earth has raised fringes? If so, it would be an answer to the question of the edges and ends of the Earth, indicating that its sides are raised, thus preventing creatures from going out of it. Such would be the case if you were to put a tiny ant in a huge container.

It would try its utmost to reach the top to get out of the container, but it would always fail to do so. We know that when these verses where sent down, Arabs, like other nations, thought the Earth is flat. Therefore, human beings could understand from this verse that the Earth has ends that are raised higher than its center. Importantly, where are its ends, and what are its boundaries? First and foremost, Allah only knows that. As a matter of fact, when I pondered and reflected on that, I came up with many associations to the statement of Allah, Almighty, in the chapter of Al-Rūm, **"In the nearest land"** [Al-Rūm:3]. Dear readers, if you were of those people who make connections between things, reading these words probably made your hair stand on end out of astonishment. Nonetheless, the verse has no contradiction, whether we arguably accept that the Earth is flat or spherical, so kindly mind that!

Allah, Almighty, said, **"Indeed, those whom the angels take [in death] while wronging themselves—[the angels] will say, 'In what [condition] were you?' They will say, 'We were oppressed in the land.' The angels will say, 'Was not the earth of Allah spacious [enough] for you to emigrate therein?' For those, their refuge is Hell—and evil it is as a destination"** [Al-Nisā':97].

His saying **"the earth of Allah spacious"** signifies that it is extensive, vast, and large in size and scale. While this does not necessarily mean that the Earth is not spherical, it better fits the flat space especially if it has ends. For instance, we say a spacious seating area, a spacious house, or a spacious garden. I have no clue who said that a spherical ball can be spacious. Anyhow, to be objective and fair to the contenders at least from my point of view, it is worth mentioning that the verse is no contradiction to the notion that the earth is spherical.

Allah, Almighty, said, **"And to Allah belongs whatever is in the heavens and whatever is on the earth. And We have instructed those who were given the Scripture before you and yourselves to fear Allah. But if you disbelieve—then to Allah belongs whatever is in the heavens and whatever is on the earth. And ever is Allah Free of need and Praiseworthy"** [Al-Nisā':131].

In myriad verses in the Quran, we find that Allah, Exalted and Glorified, states, **"And to Allah belongs whatever is in the heavens and whatever** is on the earth" [Al-Nisā':131], and "**And to Allah belongs the dominion of the heavens and the earth**" [Āl-ʿImrān:189]. The Quran is replete with such verses. The chapter of Al-Nisā' alone has six similar statements. "**Have they not asked themselves one day why He says, 'to Him belongs whatever is in the heavens and the earth?'**" [Al-Baqara:116]. Is the Earth really very insignificant and a mere atom in size as modern science claims? Does Allah say, "to Allah belongs the dominion of the heavens and the earth" [Al-Baqara:106], merely because human beings know it and live on it? What if the resources on the Earth are limited and they have to live in space or else-where, as the pioneers of these modern scientific communities proclaim? If they were to reply that such was the limit of the knowledge of human beings back then, we will accept that for the sake of argument and ask them, "Were human beings back then familiar with the heavens and the Earth only?" Why did not Allah Who know all His creations say that to Him belongs whatever is in the worlds as He stated, "[All] praise is [due] to Allah, Lord of the worlds" [Al-Fātiḥa:2]? Or is it the reality that the human existence is composed of the heavens and the Earth, as well as what is above them, what is under them, and what is around them?

Allah, Almighty, said, "**They have certainly disbelieved who say that Allah is Christ, the son of Mary. Say, 'Then who could prevent Allah at all if He had intended to destroy Christ, the son of Mary, or his mother or everyone on the earth?' And to Allah belongs the dominion of the heavens and the earth and whatever is between them. He creates what He wills, and Allah is over all things compe-tent**'" [Al-Māʾida:17].

I have written tens of research papers about al-Mahdī, the antichrist, Jesus, the son of Mary (Peace be upon him), the Torah, the Gospel and the Bible. I still have not published these research papers. In all honesty, I cannot deny that this very verse was a watershed moment that made me transform and revamp my theories regarding all my research endeavors. It is no exaggeration to say that a whole new volume could be written to comment on this verse. That is why I will refrain from commenting on this verse in this book. Instead, I will let the reader unleash his imagination to thoroughly reflect upon its meaning. If Allah wills, it will open new broad horizons for better understanding the Quran, the messiah, the Earth, the heavens, and much more.

Allah, Almighty, said, "**Allah has made the *Ka'bah*, the Sacred House, standing for the people and [has sanctified] the sacred months and the sacrificial animals and the garlands [by which they are identified]. That is so you may know that Allah knows what is in the heavens and what is in the earth and that Allah is Knowing of all things**" [Al-Māʾida:97].

Till now, I still have not comprehended the relationship between the *Kaʻba* and knowing what is in the heavens and the Earth. There must be an overriding rationale behind that! Whenever I recite this verse, I readily recall the narration ascribed to the messenger (Peace and blessings be upon him) and sometimes attributed to some of his companions and successors. This narration tells that the Earth was flattened out, starting from under the *Kaʻba,* or that the *Kaʻba* was like an island on water and then it was flattened out. Interestingly, some People of the Book deem the rock in Jerusalem as the center and the middle of the earth. The difference in whether that was Mecca or the rock in Jerusalem is another point of inquiry I shall address later in a separate volume.

As far as this topic is concerned, al-Ḥassan stated, "Allah created the earth in the place where Jerusalem is now and it was like a massive rock that has smoke attached to it. He then raised the smoke up high and created the heavens from it. Then he kept the rock where it was and then spread it out to make the earth flattened out. This is in reference to the statement of Allah, Almighty, "**the heavens and the earth were a joined entity**" [Al-Anbiyāʾ:30], that is, they were attached to one another."

In his commentary *al-Majmūʻ*, he maintained that Allah created the Earth first before leveling it out. It was in one location, namely the location of Jerusalem. He then created the heavens and then spread out the Earth. He said to it, "Be flattened out this way. Be spread out that way." They reported from ʻAtāʾ who he said, "I was told that the Earth was spread out from under the *Kaʻba*." Others also said, "It was spread out from Mecca." They related that Mujāhid said, "The holy house was one thousand years before the earth and the earth was spread out from below it."

On commenting on the verse of the chapter of Al-Baqara that we discussed earlier, al-Samarqandī stated in his commentary *Bahr al-ʿUlūm*, "It might be argued that He said in other verses, '**Are you a more difficult creation or is the heaven? Allah constructed it. He raised its ceiling and proportioned it. And He darkened its night and extracted its brightness. And after that He spread the earth**' [Al-Nāziʿāt:27–30], and He mentioned in that verse that the Earth was created after the sky, but in these verses, He stated that 'the earth was created before the sky.' The answer is that is as follows: He created the Earth as a red, rocky plateau in the location of the *Kaʿba*. After creating the sky, He spread out the Earth following the creation of the sky. This is in reference to His saying, '**And after that He spread the earth**' [Al-Nāziʿāt:30], which means He flattened it out."

The crux of this discussion is that the Earth was created before the sky as a red plateau in the location of the *Kaʿba*. On the whole, the first conclusion is that the creation of the Earth preceded that of the sky, whereas the second conclusion is that the beginning of the Earth is the location of the *Kaʿba*. Speaking of the *Kaʿba*, I bet you, dear readers, know the well-known hadith attributed to the messenger (Peace and blessings be upon him) and to a group of his companions regarding the sacred frequented House, don't you? It is the mosque that is in the sky or the *Kaʿba* that is in the sky which is so aligned with the *Kaʿba* that is on the Earth that if it were to fall, it would fall on the actual *Kaʿba*. This hadith is equally known by the lay people as well as the learned ones.[1] If we accept the hadith, it serves as evidence to substantiate the fact that the Earth neither moves nor revolves but rather is firm, static, and stable.

---

1 ʿAbdullah ibn ʿAbbās narrated an account that he attributed to the Prophet (Peace and blessings be upon him), in which he stated: "the sacred frequented House in the sky is dubbed *al-Ḍarḥū*. It is perfectly aligned with the Holy Mosque. If the former were to fall, it would fall right on top of it. Every day, seventy thousand angels get into it and they never see it. It has a sanctity in the sky as that on Mecca [on earth]." This account was recorded by ʿAbdel-Razzāq in his *Muṣannaf* hadith collection (8874) and al-Ṭabarānī in his collection *al-Muʿjam al-Kabīr* (11/417) (12185), and al-Bayhaqī in his book *Shuʿab al-Īmān* (3709). Ibn Ḥajar al-ʿAsqalānī (deceased 852 AH) in his book *Tuḥfat al-Nubalāʾ* on page 80: Its chain of transmitters includes Abū Ṭufayl who is a weak narrator. To be continued.

Allah, Almighty, said, "**[All] praise is [due] to Allah, who created the heavens and the earth and made the darkness and the light. Then those who disbelieve equate [others] with their Lord**" [Al-An'ām:1]. This verse is similar in meaning to the first three texts of the Torah, which read, "In the beginning, God created the heavens and the earth. *(1) In the beginning God created the heaven and the earth. (2) And the earth was without form, and void; and darkness was upon the face of the deep. And the Spirit of God moved upon the face of the waters. (3) And God said, Let there be light: and there was light.*"

I remember that when I found such similarity between the texts, I thought I was the first to do so. Given that I was still a novice intellectually and spiritually, I was as excited and delighted as a naive child to accomplish such an insignificant task that I considered significant back then. Then, I came across the statement ascribed to Ka'b al-Aḥbār regarding this same verse. It reads, "This verse is the first verse of the Torah." The secrets of this verse and its striking manifestation and indications are numerous and significant. Perhaps I will address them on a different occasion.

What I want to say in this regard is that the explicit meaning of the Quran, like that of the Torah, is that the darkness and the light existed before the creation of the sun and the moon. Producing the darkness and the light does not mean creating them after the creation of the sun or anything like that. Nevertheless, the most preferable view is that of al-Ṭabarī, who stated, "He darkened the night and brightened the day," that is, He made the darkness the nature of the night and the brightness the nature of the day. In this regard, some Muslim scholars made interconnections between the sun and the moon with the day and the night. Yet, in my point of view, this sort of interrelation is flawed. They opined so because of the statement of Allah, Almighty, "and made" which means they are not independent or self-sustained. That is why they were created and made as such. How so? Through the sun. If it sets, the world becomes dark, and when it rises, the world becomes bright. Still, there is a blatant flaw is such an argument, as I have told you. The sun has nothing to do with light. Instead, it is the moon only, and whoever look at the moonlight and the sunshine will notice the difference.

It seems to me that the darkness and light in this verse mean guidance and misguidance. This is what some commentators concluded. Al-Ālūsī (May Allah have mercy on him) recorded their view regarding that as follows: "What is meant by darkness is misguidance which has various forms, whereas what is meant by light is guidance which is only one. The following statement of Allah, Almighty, **'And, [moreover], this is My path, which is straight, so follow it; and do not follow [other] ways, for you will be separated from His way. This has He instructed you that you may become righteous'** [Al-Anʿām:153], signifies diversity and multiplicity. On the other hand, more than one commentator were of the opinion that the darkness and the light actually refer to the tangible darkness and light. In this verse, this meaning is also plausible, for the term should be first and foremost refer its explicit reality. Here this meaning is conceivable on the grounds that the terms have plausible meanings that they explicitly denote because they are associated with the heavens and the earth."

Al-Ālūsī's words indicate that the intended meaning of this verse is that it primarily refers to the darkness of misguidance and the light of guidance, even if we were to arguably agree that the darkness and the light could also mean the tangible natural agents of the day and the night. You can consult the chapter of *The Day and the Night* to find that this very verse has nothing to do with the sun and the moon, but it is associated to the essence of the day and night. In brief, those who argued the day and the night are manifestations of the sun and the moon have no single verse in the Quran to corroborate their view. Even worse, this and other Quranic verses clearly reveal that the darkness and the light existed before the sun and the moon, exactly as is mentioned in the Torah. As for the issue of whether the day and the night are accidental or essential, we shall discuss that in a separate chapter, God willing.

Al-Fakhr al-Rāzī stated in his commentary, "the darkness precedes the light in terms of occurrence and realization. That is why it should be mentioned first in utterance. What could substantiate this view is the narrated accounts from Allah that He, Almighty, created the creations in darkness and then He showered them with His light. The fourth point: One might argue that why the darkness is mentioned in the plural form in Arabic while the light is in the singular? In response, we say: If one understands darkness to refer to disbelief and light to refer to the truth, then the meaning is crystal clear since disbelief comes in different forms whereas belief is just one. As for those who understand them to mean tangible darkness and light, our response is that the light is an epitome of such completely perfect form of brightness. Such brightness can be diminished gradually giving rise to different degrees of darkness. Hence, the term darkness was mentioned in the plural form."

Allah, Almighty, said, **"Say, 'Is it other than Allah I should take as a protector, Creator of the heavens and the earth, while it is He who feeds and is not fed?' Say, [O Muhammad], 'Indeed, I have been commanded to be the first [among you] who submit' [to Allah] and [was commanded], 'Do not ever be of the polytheists'"** [Al-Anʿām:14]. Regarding His saying **"Creator of the heavens and the earth,"** if we adhere to the classical commentaries that state that He created them and made them from scratch in an unprecedented form, then this is a counterargument to those who promulgate today the hypothesis of multiple universes, arguing that the creation of this universe occurred millions of times—if not ad infinitum—in the past until this universe of ours came into being in the form we know it now.

In response, I argue that besides countering their argument, His saying **"Creator of the heavens and the earth"** could also mean that He is the One Who separated and split them. We say that the heavens and the Earth were once one mass, or a single bubble, and they were jointly attached. He then created them in the sense that He separated and detached them, making the heaven up high and leaving the Earth in the bottom. You should also reflect on the phrases that people repeatedly use such as the fasting person broke his fast (*faṭara*) or his heart was broken (*fuṭira*). The same goes for His saying, **"[And] who created seven heavens in layers. You do not see in the creation of the Most Merciful any inconsistency. So return [your] vision [to the sky]; do you see any breaks?"** [Al-Mulk:3]. The term *fuṭūr* means breaks or flaws. You should survey all the verses and examples in which the terms *fāṭir* or *faṭara*, and you will realize—if Allah wills—that the view I espouse regarding the meaning of the verse is the soundest in meaning. Moreover, the meaning I detailed earlier is perfectly consistent with what is mentioned in the Torah.

Allah, Almighty, said, "**And if their evasion is difficult for you, then if you are able to seek a tunnel into the earth or a stairway into the sky to bring them a sign, [then do so]. But if Allah had willed, He would have united them upon guidance. So never be of the ignorant**" [Al-Anʿām:35]. It is true that this verse does not decisively indicate that the Earth is flat, but it suggests to the reader that it could be flat and not spherical. Have you ever heard about a tunnel in a spherical ball or a circular shape? This, however, could truly be the case in flat objects.

Allah, Almighty, said, "**And with Him are the keys of the unseen; none knows them except Him. And He knows what is on the land and in the sea. Not a leaf falls but that He knows it. And no grain is there within the darknesses of the earth and no moist or dry [thing] but that it is [written] in a clear record**" [Al-Anʿām:59]. Have we ever wondered what is meant by the darknesses of the Earth? And why did Allah mention the land, the sea, and then the darknesses? Is not the Earth today anything more than the land, the sea, and what is inside the Earth? It is known that the phrase "**what is on the land**" is understood to mean that which is on it or in it. The same goes for the sea. Do you not argue that Google Maps shows every nook and cranny of the earth? Do you not further argue that the sun rises on the Earth, the moon, the stars, and so on? If so, where is the piece of earth that is dark all the time? Its depths, right? Or do you mean something greater than that, such as what is beneath the Earth? I mean the darknesses that are beneath the Earth, namely what is called "the Deep" in the holy Bible. Thankfully, I am not the only one whose mind went very far reflecting on that. Some Quran commentators said, "the darknesses of the Earth refer to the rock which is beneath the seventh earth."

There is a peculiar narrative on this verse in the commentary of Ibn Kathīr as follows: "As for His saying '**And no grain is there within the darknesses of the earth and no moist or dry [thing] but that it is [written] in a clear record**' [Al-Anʿām:59], Muḥammad ibn Isḥāq narrated on the authority of Yaḥya ibn al-Naḍr who related from his father who heard ʿAbdullāh ibn ʿAmr ibn al-Alʿāṣ as saying: There are too many jinn below the third earth and above the fourth earth that if they were to look at you, you would never be able to see light again. There is a seal of the seals of Allah, Exalted and Glorified, at every corner of the earth. There is also an angel on every seal and, every day, Allah, Exalted and Glorified, dispatches to that angel another angel of His saying: Keep what you have." Putting the authenticity of the narration aside, it still encompasses a great deal of evidence that the primordial stance that transmitters espoused at the time exquisitely corresponds to the position that the Earth is flat and outrightly contradicts the position that it is spherical.

Allah, Almighty, said, **"And if you obey most of those upon the earth, they will mislead you from the way of Allah. They follow nothing but speculation; they are merely guessing"** [Al-An'ām:59]. This verse is primarily a counterargument to those who contend that the agreement of the vast majority of people substantiates the soundness of their hypotheses and theories. When you discuss with people such topics as the shape of the Earth and even religious matters, many of them are quick to say that it is not conceivable that you and some obscure figures are right, and the millions of people are wrong! Definitely, you are plagued by arrogance. Then, they hurl the worst labels at you. In actuality, I even saw the most educated and open-minded amongst them follow suit. They derive their strength from such blind, majoritarian approval. This verse is a decisive proof that undermines their notion that flagrantly contradicts common sense.

As a matter of fact, the bulk of people jump on the bandwagon in a blind imitation of others and that is why they are big in number. If a distinguished dignitary from amongst them held a certain view, they mindlessly imitate him or her out of fear or respect for that figure. As such, their number dramatically multiplies manifolds. On the contrary, truth seekers often find themselves in a direct confrontation of rote learning and what people have long taken for granted. Therefore, they need to have a great deal of patience. Hence, we understand why Allah, Exalted and Glorified, stated this sublime verse in His Glorious Scripture in order for it to last as long as the Quran lasts in this mundane world. Not only does this verse fuel the patience of the truth seekers and the promulgators of it. but it also serves as a retort to those who derive their validity and legitimacy from majoritarian mindless imitation. As Allah, Exalted and Glorified, portrayed them, they **"follow nothing but speculation; they are merely guessing"** [Al-An'ām:116].

Allah, Almighty, said, "**And it is He who has made you successors upon the earth and has raised some of you above others in degrees [of rank] that He may try you through what He has given you. Indeed, your Lord is swift in penalty; but indeed, He is Forgiving and Merciful**" [Al-An'ām:165]. He also said, "**And We have certainly established you upon the earth and made for you therein ways of livelihood. Little are you grateful**" [Al-A'rāf:10]. Why did not Allah say, "We have established you on the Earth and the heavens and made you successors on the heavens?" Just imagine with me if people were to live on other planets, as futurist modernists claim that people will live on other planets in the future, how would they feel when they read these verses? Their only response would be that Allah was addressing the people at the time, and that this verse does not speak to them, and its meaning is clear. However, for those who really understand and know that what they predict will never take place in the future, their faith will increase every passing day as they read this verse. Whenever they recite this verse, they know that the predictions of the scientific community will never occur. Then, they will follow it by reciting, "**they follow nothing but speculation; they are merely guessing**" [Al-An'ām:116]. Then, their hearts and minds will be more inclined to better understand His saying, "**Say, 'With Allah is the far-reaching argument. If He had willed, He would have guided you all'**" [Al-An'ām:149].

Allah, Almighty, said, "**Do they not look into the realm of the heavens and the earth and everything that Allah has created and [think] that perhaps their appointed time has come near? So in what statement hereafter will they believe?**" [Al-A'rāf:185]. How were human beings able to look into the realm of the heavens and the Earth at the time of revelation of the Quran against the backdrop of the modern definition of the heavens, that is, if the heavens were above space? However, if we were to consider that the existence, or the universe, as we technically call it today, then human beings in the past would be able to look into it. Why did not He say "universes?" What is the position of those who claim that they look at virtual dimensions?

Allah, Almighty, said, "**Allah has already given you victory in many regions and [even] on the day of Hunayn, when your great number**

pleased you, but it did not avail you at all, and the earth was confining for you despite its vastness; then you turned back, fleeing"
[Al-Tawba:25].

His saying "**despite its vastness**" indicates that they felt confined, in spite
of the fact that the Earth was very spacious. Human beings typically feel confined when they get into something that restricts them from either two or four
directions. It is nearly impossible to say that this is true if it occurs from just one
direction. In the standpoint of modern science today, enclosure could apply to
the spherical shape from one direction.

Let us think about such phrases as: their selves were confining them; a tight
garment; and narrow-mindedness. It could be inferred from this verse that the
Earth is somehow encircling people in it. In other words, it has raised ends and
edges that are higher than creations. This is in line with His saying, "**O company of jinn and mankind, if you are able to pass beyond the regions
of the heavens and the earth, then pass. You will not pass except
by authority [from Allah]**" [Al-Raḥmān:33]. This verse further provides a
counterargument to those mocking us who think the Earth we refer to is a flat
earth that floats in space. They allege that the water overflows from the Earth
into space. Nonetheless, had they accepted that the Earth has ends and fringes
that are raised higher than the water level, they would not have fallen in the trap
of such a fallacious claim in the first place.

Allah, Almighty, said, "**Indeed, in the alternation of the night and
the day and [in] what Allah has created in the heavens and the earth
are signs for a people who fear Allah**" [Yūnus:6].

Is that which is high above us is called heavens? And is it possible to conquer the space? The most salient question is, "Does the atheist fear Allah?
Where will he find the signs, if he is not mindful of Allah?" The other issue
is that human beings can look into what is in the heavens and that it is not
unknown to them, as it is the case nowadays, as space agencies say. Primordial
human beings managed to look into the seven planets, the stars, and so forth,
and this continued even before the advent of the religion of Islam. But where
are the seven heavens today, as per those who mindlessly believe in space
agencies?

Allah, Almighty, said, "**And it is He who created the heavens and the earth in six days—and His Throne had been upon water—that He might test you as to which of you is best in deed. But if you say, 'Indeed, you are resurrected after death,' those who disbelieve will surely say, 'This is not but obvious magic'**" [Hūd:7]. What is the Throne? Is it like the throne on which kings sit and rest? Or is it the ceiling of the creations? Or is it a tangible being that conjures up images of awe, supremacy, and control as some contemporary commentators argued? Or is it the mechanism thereby divine decrees are communicated to the universe? Or is it the remote control of the creations? Or it is the order of the universe that works like the arteries and nerves of the human body? These are the views posited by scholars, past and present. Still, I have not come across anybody who maintain the same views and thoughts as those I have in mind. Nonetheless, I still think these thoughts are far from accurate. What if the throne is the sum of the heavens and the earth; I mean, the bubble of the universe. If this is not the case, then can it be that which is above the seven heavens? What if it means the entirety of the Earth? There are myriad questions, imaginations, and flights of fancy that I am not sure if I will be able to jot down or at least write about the ones I have studied and the conclusion I have drawn from them.

Allah, Exalted and Glorified, said, "**Have they not seen that We set upon the land, reducing it from its borders? And Allah decides; there is no adjuster of His decision. And He is swift in account**" [Al-Ra'd:41]. He, Almighty, also said, "**Then do they not see that We set upon the land, reducing it from its borders?**" [Al-'Anbiyā':44]

Where are the borders of the Earth? The context of the verse shows there is a threat for disbelievers that their land will be stripped off of them or will be ruined, which is historically factual. I know this very well. However, I want to say that Allah has stated that the Earth has borders, and perhaps it is referring to the lands of disbelievers. It is known that all potential meanings the verse propounds are taken into consideration. In other words, if there are multiple facts and complementary meanings potentially suggested by the verse, they are to be considered.

Thankfully, this "borders" notion lines up with the perspective of the flatness of the Earth. Should the Earth be flat, it would necessarily have borders, as opposed to the spherical Earth. Following the former interpretation, multiple meanings [of the verse] will be produced whereas the latter interpretation produces only a single meaning, which is the lack of the disbeliever's domination over the land. Still, this interpretation will leave ample room for questions. Al-Sha'by is reported to have been said, "If the Earth [literally] reduces, your garden will become narrow, but it is the fruits and lives that are subjected to reduction." This meaning is valid. It can be said, however, that the mind of al-Sha'by had not fully grasped that the Earth is spacious. Allah said, **"And it is He who spread the earth."** When Allah, Exalted and Glorified, says that the Earth is spread, nobody knows the borders of this spreading except Him, Almighty. One relevant narration to the above is attributed to 'Ikrima who said, "Were the Earth to reduce, we would not find a place to sit on." It shares the same response as the previous.

Allah, Almighty, said, **"It is Allah who created the heavens and the earth and sent down rain from the sky and produced thereby some fruits as provision for you and subjected for you the ships to sail through the sea by His command and subjected for you the rivers"** [Ibrāhīm:32].

I have a lot to say, but I will keep it to myself. Just, let us ask ourselves with reflection, "Why would Allah, Exalted and Glorified, say that He subjected the ships for us though humans are the ones who created it? Is it not just enough to say that the human's action is created by Allah or inspired by Him? What if someone approaches you saying that it was Satan who inspired me or anything in those lines? How would you refute his argument beyond recovery?" This verse requires a deliberate process of reflection. I will delay explaining my hypothesis for few years, though I have inferred it here to provoke the mind of the reader to research it himself.

Allah, Almighty, said, **"And He has cast into the earth firmly set mountains, lest it shift with you, and [made] rivers and roads, that you may be guided"** [Al-Naḥl:15]. In another verse, Allah, Almighty, said, **"And We placed within the earth firmly set mountains, lest it should**

shift with them, and We made therein [mountain] passes [as] roads that they might be guided" [Al-'Anbiyā':31].

Reflecting on those verses and similar ones reveals that mountains were created and set to firm the Earth and prevent it from shifting. This implies that the Earth was shaking and moving, thus inhabitable. One may ask, "What is the secret behind the movement it used to make? Is it an intrinsic movement like the earthquakes?" Assuming it is earthquakes, we do witness today the occurrence of earthquakes across the world. Today, the scientific society maintains that the Earth is orbiting in the sky with a speed of a thousand miles per hour. Let us imagine this intangible velocity! I would ask whether this velocity was proportionally higher before the mountains were set, or is it faster today with an intangible speed no human being can feel? How strange is this? How could people believe in such theories when it is hard for them to believe in Allah?

My viewpoint is that the Earth was initially above water with a flat surface. When you throw a wooden bar on top of the water, what happens to it? It moves right and left, and any slight push will shake it off. However, if you place weights on both edges, what happens? The wooden bar will move steadily, nor randomly. I believe this is exactly what happened. Allah set the mountains like giant wedges that penetrated the Earth to firm its movement. It does not move today, except in the instances of earthquakes. The Earth itself does not move, but rather the mountains; or, in better words, pass like clouds. If you can imagine this, you will find that all the Quranic verses line up seamlessly and provide a realistic interpretation comprehended by the aged, the young, the scientist, and the layman.

Evidence of such a point of view can be found in the reports. I shall provide some of them as follows. Ibn Jarīr commented, "Allah, Exalted and Glorified, imparted firmness to the Earth through the mountains lest it shakes His creation who inhabit its surface. It was indeed shaking before mountains were set above it. It was narrated to us by Bishr from Yazīd from Sa'īd from Qatāda from al-Hasan from Qays ibn Ubāda who said that the Earth was shaking violently when Allah, Exalted and Glorified, first created it. The angels said, 'This would not provide any stability for anyone above it.'

"Thereafter it became stable with its own firm mountains. Al-Muthanna narrated to me from al-Ḥajjāj ibn al-Minḥāl from Ḥammād from ʿAtāʾ ibn al-Sāʾib from Abd Allah ibn Ḥabīb from Ali ibn Abu Ṭālib who said, "After Allah created the Earth, it was shaky and complained, 'My Lord, why will you place the children of Adam above me to commit major sins and spread impurity above me? Allah then set firm mountains above it, some of which are observable while others are not. The bed of the Earth was shaky.' *Mayd* expresses imbalance and swaying. When the word is collocated with a ship, it means the ship is swaying with its passengers. The word *mayd* also refers to the seasickness. Interpreters of the Quran have made similar statements. For example, al-Muthanna narrated from Abu Ḥudhayfa from Shibl from ibn Abu Najīḥ from Mujāhid who commented on 'lest it shift with you' as 'it sways with you.' A similar narration was narrated by al-Qāsim from al-Ḥusayn from Ḥajjāj from ibn Jurayj from Mujāhid.

"Al-Ḥasan ibn Yaḥya said that Abd al-Razzāq said that Maʿmar narrated to us from Qatāda from al-Ḥasan commented on the saying of Allah, '**And He has cast into the Earth firmly set mountains, lest it shift with you, and [made] rivers and roads, that you may be guided**,' by saying, '[Allah] set mountains to prevent the Earth from shaking.' Qatāda said that he heard al-Ḥasan saying, 'The Earth was shaking after Allah created it.' They said, 'This [Earth] would not provide any stability for anyone above it. In the morning, the mountains were created but the angels had no clue from what the mountains were created.'"

Allah, Almighty, said, "**[Iblees] said, 'My Lord, because You have put me in error, I will surely make [disobedience] attractive to them on earth, and I will mislead them all'**" [Al-Ḥijr:39].

In this regard, the following question is in order: "Would Satan and his attempts of misguidance cease to exist were we to go to the space? If not, why would he say, '**I will surely make [disobedience] attractive to them on earth**?' How about the humans' dreams of building settlements on Mars? Would those who travel to space and live there instead of the Earth be outside the reach of Satan? Or would we say that this verse addresses those who lived during that period exclusively?" Statements like "the point is the purpose not the details" are made by those who escape from any serious attempt to answer those questions.

Allah, Almighty, said, "**They ask you, [O Muhammad], about the Hour: when is its arrival? Say, 'Its knowledge is only with my Lord. None will reveal its time except Him. It lays heavily upon the heavens and the earth. It will not come upon you except unexpectedly.' They ask you as if you are familiar with it. Say, 'Its knowledge is only with Allah, but most of the people do not know'**" [Al-Aʿrāf:187].

Do the heavens and the Earth represent the whole existence? Where else does it lay heavily? Had it not been laid heavily in space or those new Earth-similar planets they discovered? Or do the heavens and the Earth comprise all existence, otherwise currently called the Cosmos, as I mentioned in the earlier chapter of the heavens and the Earth? Certainly, this verse not only refutes the proponents of the sphericalness of the Earth but also refutes many modern-day imitators, though in a different context than the one we are currently addressing.

Allah, Almighty, said, "**I did not make them witness to the creation of the heavens and the earth or to the creation of themselves, and I would not have taken the misguiders as assistants**" [Al-Kahf:51].

How amazing this is? This verse represents a prophecy for the people who mislead the people away from the cosmic facts. Consider how they interpret the inception of the universe and make movies as if they were witnessing the beginning of creation alongside Allah, Exalted, He above such thing. They mislead people. For this reason, anyone who desires to learn about the creation of the heavens and the Earth must consider the Quran, for it sufficiently expounds the manner, period, elements, creatures, and the order [of creation].

Allah, Almighty, said, "**A revelation from He who created the earth and highest heavens**" [Taha:4]

Why did Allah specify the Earth and the highest heavens? There are many reasons, most important of which is that the drama of human existence is definitely implied. Why did Allah begin with the Earth, though most of the Quranic verses begin with the mention of the heavens followed by the Earth? Some commentators say it is for the purpose of creating harmony between the verses. Others say it is a completion of what is mentioned earlier. Since the addressee is the Messenger (peace and blessings be upon him) who resides on Earth, Allah started with what is closer to him. One more opinion maintains that it is because of formation. It is said Allah began with the lower, then the higher, or began with the weaker, then the stronger.

However, I think the reason for the initial mention of the Earth before the heavens is because the Earth is actually created before the highest heavens, as proved by the literal meaning of the Quranic verses. It can be said that the heavens retain a higher and more prestigious status than the Earth, though the latter is existentially older. Dear reader, pay attention that this instance is mentioned only in the chapter of Ṭaha. In this particular chapter, Aaron was mentioned before Moses. Allah said, **"So the magicians fell down in prostration. They said, 'We have believed in the Lord of Aaron and Moses'"** [Ṭaha:70]. In the chapter of al-Shuʿarāʾ, Moses is mentioned before Aaron, **"The lord of Moses and Aaron"** [al-Shuʿarāʾ:48].

It is popularly known that Moses is greater than Aaron, because the former is one of the strongly determined messengers, so what is the reason? The reason is different from the one mentioned in exegeses' books. In my opinion, Aaron is the older brother of Moses, which is mentioned in the Book of Exodus in the Torah. Aaron, being the older one, was given precedence just as the Earth came first because it is created first. Since the highest heavens retain a seemingly higher status than the Earth, we find in this very chapter the saying of Allah, Exalted and Glorified, to Moses after fearing what the magicians threw, **"Allah said, 'Fear not. Indeed, it is you who are superior'"** [Ṭaha:68].

Just as the heavens outclass the Earth, Moses is superior as far as status. Perhaps the initial mention of [Aaron] is that the young Moses sensed apprehension in overtaking him, in addition to the belief of the magicians. As a matter of fact, I have more to say about the reason for the initial mention of Aaron, but it would be understandable to a very little number of people like mathematicians and Sufis, though I do not claim I am one of them. I still think they would understand. However, I will refrain from indulging this point, lest the reader become distracted from the overall purpose of the book.

Allah, Almighty, said, **"To Him belongs what is in the heavens and what is on the earth and what is between them and what is under the soil"** [Ṭaha:6]

In this verse, Allah, Exalted and Glorified, mentioned the heavens and the Earth and what is between them, thus including everything. But, why did Allah add **"and what is under the soil**?" Is water or anything else under the Earth according to my hypostasis? Is there more ancient evidence that the soil may be something under the Earth or just under and inside the land? Linguistically, soil sometimes refers to the wet dust and other times to anything wet. Let us explore the statements of the commentators:

Al-Tabary (may Allah have mercy upon him) said after relating what he narrated from al-Daḥḥāk, "What is meant here is what is underneath the seven earths." Al-Zamakhshary (may Allah have mercy upon him) said, "It refers to what is under the seven earths." Muḥammad ibn Kaʿb said, "It refers to what is under the seventh earth." Al-Rāzī, whom they always refer to as an example of the early scholars who supported the sphericalness of the Earth, said, "Assuming the soil is the last surface of the world, then nothing would be underneath it, but how would He own it? We say that the linguistic meaning of soil is the wet dust, which makes it likely to have something underneath it; either the bull, the whale, the rock, the sea, or the air according to the many narrations on this regard." Very similar statements are found in the exegeses of Shi'a, Sufism, Ibadism, and Zaydism.

Allah, Almighty, said, **"[It is He] who has made for you the earth as a bed [spread out] and inserted therein for you roadways and sent down from the sky, rain and produced thereby categories of various plants"** [Taḥa:53]. And, **"[The one] who has made for you the earth a bed and made for you upon it roads that you might be guided"** [Al-Zukhruf:10]. And, **"Have We not made the earth a bed (i.e. resting place]?"** [Al-Naba':6].

The meaning of Allah's, Exalted and Glorified, saying, **"the earth a bed"** is void of any implication to the sphericalness of the Earth. It leans toward suggesting the flatness of the Earth, because there is not even a distant linguistic reference to any sort of sphericalness in the word "bed." Al-Qurtuby said, "It is a bed and an expanse." Ibn Kathīr said, "It means that the Earth is an expanse and firm above which you walk, stand, sleep, and act though it was created above the wavery water. Allah firmed it with mountains lest if shifts sideways."

In *Mafātīḥ al-Ghayb*, al-Rāzī said, "We mentioned earlier even though the Earth is a bed, it is as such because it is standing still. It retains certain characteristics that makes it useful for agriculture and erecting buildings to cover the errors of the alive and the dead. And since the bed is the resting place of the child, the Earth is made as a bed because of the abundant resting places it provides." Al-Rāzī, as well as the other commentators and linguistics, understood that the Earth is still and motionless from the fact that it is a bed. Have you ever tried to put an ant on giant ball and then spin it around its axis at a thousand miles per hour? Could this ball serve as a bed for that ant? Qatada commented on this verse, "The earth is an expanse." Ibn Kathīr said, "It is paved for the creatures and it is firm and still." Al-Zamakhshary said, "It is a bed." A similar statement is made by al-Bughawy.

Allah, Almighty, said, **"Blessed is He who has placed in the sky great stars and placed therein a [burning] lamp and luminous moon"** [Al-Furqān:61].

One day, as I was reciting this verse, I wondered, "How could they say there are more than two billion suns apart from ours in this cosmos, plus the 193 moons present in our solar system alone while the overall number of moons is uncountable?" Allah, Exalted and Glorified, said on the other hand, **"He who has placed in the sky great stars and placed therein a [burning] lamp and luminous moon."** Were there more than two billion suns—and Allah knows best what He created—He would have made a reference to them in the Quran. The fact is, nevertheless, the entire Quran from the beginning till the end informs of just a single sun and a single moon, which is something of a commonality among religions. Even at fairytales and myths, they couple the sun and the moon as twin gods, or as a brother and a sister like Apollo and Artemis. Furthermore, it seems there is only one sun and one moon in the Quranic verses mentioning the heavens Allah created. The proof can be found in the verse in chapter Nūḥ that reads, **"Do you not consider how Allah has created seven heavens in layers. And made the moon therein a [reflected] light and made the sun a burning lamp?"** [Nūḥ:15–16]. The sun and the moon are at the heavens. This refutes who maintains there are billions of suns and moons. In Chapter al-Qiyāma, **"And the sun and the** moon are joined. Man will say on that Day, "Where is the [place of] escape?" [Qiyāma:9–10]. In the Hadith, the sun and the moon are folded up. Allah, Exalted and Glorified, said, "Do you not see that to Allah prostrates whoever is in the heavens and whoever is on the earth and the sun, the moon, the stars, the mountains, the trees, the moving creatures and many of the people? But upon many the punishment has been justified. And he whom Allah humiliates—for him there is no bestower of honor. Indeed, Allah does what He wills" [Al-Ḥaj:18]

Reflect on those verses that mention no suns or moons other than the single sun and moon we have. Were there other suns and moons, the Quran would state it as well as the Prophetic Sunnah and the reports attributed to the companions and successors.

Allah, Almighty, said, "To Him belongs whoever is in the heavens and the earth. And those near Him are not prevented by arrogance from His worship, nor do they tire" [Al-Anbiyā':19].

Have you ever thought why Allah said, "and those near Him?" Allah preceded it by saying that everything in the heavens and the Earth belong to Him. This includes the worldly human existence and other creatures. He mentioned the locations, namely, the heavens and the Earth. Now, what about "those who are near Him?" Are they within the heavens and the Earth, or beyond them? Or is it just to boost stature? I personally rule out the last probability and think they are beyond the heavens and the Earth, where the angels and other creatures live. Also, I think the referents here have free choice and are susceptible to weariness, which is why Allah praised them with lack of arrogance and weariness, otherwise they would innately be forced to do so, thus making the praise unjustifiable. Allah knows best about their essence.

At any rate, the reference to an existence other than the heavens and the Earth implies it is beyond them. It is possible to locate it in the sky, as in Allah's, Almighty, saying, "Do you feel secure that He who [holds authority] in the heaven would not cause the earth to swallow you and suddenly it would sway?" [Al-Mulk:16]. Lastly, it should be highlighted that [Allah], "creates that which you do not know" [Al-Naḥl:8]; and, "And none knows the soldiers of your Lord except Him. And mention of the Fire is not but a reminder to humanity" [Al-Muddathir:31].

Allah, Almighty, said, "**And the heaven He raised and imposed the balance, that you not transgress within the balance. And establish weight in justice and do not make deficient the balance. And the earth He laid [out] for the creatures**" [Al-Raḥmān:7–10]

Reflect on the parallelism between the verse, "**And the heaven He raised and imposed the balance,**" and, "**And the earth He laid [out] for the creatures.**" The heaven is raised while the Earth is laid. It is a parallel between what appear to be two opposites. The only thing we can understand in this context is that the Earth is essentially different from the heavens and lying underneath. It is not roaming in space or a particle's size in comparison to its respective galaxy. The Earth here is parallel to the heaven. Does the word "laid" mean the same as "raise?" The first thing that comes to mind is that He lowered it and made it stable, firm, and habitable for the creatures.

For this reason, we find in ibn Kathīr's commentary, "Just as Allah raised the heaven, He laid out the Earth and made it a resting place by setting massive and lofty mountains to provide stability for the all types, shapes, colors, and languages across the Earth." In the commentary of al-Zamakhshary, he said, "Allah lowered [the Earth] and spread it above the water." Al-Rāzī said, "The Earth is laid for all who inhabited it." The Shi'i commentator al-Ṭabṭabā'y said in his commentary 'al-Mīzān, "There is an apparent remark in paralleling between the laying out the Earth and raising the heaven high." Al-Māwardy said, "Allah spread it and paved it for the creatures to fee stability above it and seek their provision."

In relation to Allah's, Exalted and Glorified, saying, "**And the heaven He raised,**" how would we understand the raising of the heaven if we were to follow the perspective that the heaven is encapsulating the Earth? Above what did Allah raise the heaven? What is underneath it? Is it possible to say the mother's belly is raised from the fetus inside it? Nobody says that, because the belly is hosting and encapsulating the fetus. In contrast, if you were to compare between the Earth and the heaven, the meaning of Allah's, Exalted and Glorified, saying, "**And the heaven He raised**" would line up perfectly without having to mention the Earth. This is understandable by anyone who comprehends Arabic language.

In Lubāb al-Tā'wīl, al-Khāzin said, "The heaven is raised above the Earth. On this regard, al-Biqāʿī provided some good remarks in his book Naẓm al-durar fī tanāsub al-āyāt wa-al-suwar. He said, 'He raised the heaven after it was attached to the Earth. He clefted them and raised it above. Evidently, any rationale man realizes that any heavy object requires an actor for raising it, Who, in this case, is Allah, because He is only One capable of such effect. Due to the high status of heaven, He fronted it before the verb to underscore His glorified creation. Allah made the heaven the first destination for His decrees and judgment, the receptor of His commands and prohibitions, the host of His angels who are sent down with revelation upon His prophets.' He continued, 'After Allah brought attention to His blessings indicating His power to raise the heaven, He mentioned, in the same manner, its parallel while highlighting the justice involved in their creation by mediating the [word balance] to show attention paid to it. **'And the Earth'** means [Allah] laid out the Earth. This structure gives the impression that the doer is mentioned twice to manifest how great Allah's disposal of affairs. 'Laid out' means He spread it above water for all creatures after imposing His balance without which the Earth cannot survive."

17

REFUTATION OF THE ALLEGEDLY QURAN-SUPPORTED CLAIMS THAT THE EARTH IS SPHERICAL

I N THIS CHAPTER, I shall refute the evidence provided by the proponents that the Earth is spherical. They use Quranic verses. Dear reader, let us consider some of their evidence.

"And after that He spread the earth" [Al-Nāziʿāt:30].

Some contemporaries cited this verse as a proof that the Earth has an oval shape or spherical. They claim that the Arabic equivalent of spread *Daḥa* means "made it like an egg or the ostrich's egg." Interestingly, I have not found any proof supporting this claim in any of the language lexicons, so I do not have the fairest clue where they found this imagined meaning. Most of the Quran commentators have interpreted the Arabic word *Daḥa* as spread. Others explained it in the light of its following verses, **"He extracted from it its water and its pasture"** [Al-Nāziʿāt:31]. So, where is the meaning "oval" coming from? Let us explore the statements of the commentators:

Ibn ʿAbbās said that *Daḥa* means "spread it." Ibn Zayd said, "He clefted and filled it with crops." Ibn Jarīr al-Tabary said, "*Daḥw* means to spread in the language of the Arabs. Interpreters have made similar statements. Bishr narrated from Yazīd from Saʿīd from Qatāda who said that *Daḥa* means to spread.

Muḥammad ibn Khalaf narrated to me from Rawwād from Abu ḥamza from al-Suddy who said that *Daḥa* means to spread. Ibn Zayd said on this regard what Yūnus related to me. He said ibn Wahb narrated from ibn Zayd that *Daḥa* means to clefted and filled it with crops."

He recited, "He extracted from it its water and its pasture;" and he also recited, **"Then We broke open the earth, splitting [it with sprouts]"** ['Abasa:26], until, **"And fruit and grass"** ['Abasa:31]. After Allah clefted it, He extracted crops out of it, and then he recited, **"And [by] the earth which cracks open"** [Al-Ṭāriq:12].

Al-Qurtuby said, "*Daḥa* means to spread." He continued, "Arabs use this word to express the meaning of spreading above the face of the Earth." Al-Zamakhshary said, "Allah spread and made it a resting place for inhabitation. Then He explained the means by which rest is acquired such as providing food, drink, and stability by extracting water and pasture and setting mountains as firm wedges to provide it with stability." In the commentary of al-Fayrūzābādī, he said, "Allah spread it above water. It is said He did so after two thousand years."

Al-Rāzī said, "There are several points here. First: *Daḥa* means to spread." Zayd ibn 'Amr ibn Nufayl said, "Allah spread it steadily above the water and then set the mountains above it. In the hadith of 'Ali (may Allah be pleased with him), he narrated, 'O Allah, the One who spread (*Daḥa*) the seven earths.' It is said that the etymological meaning of the word of *Daḥa* is to transport something from a location to another. When the word is used in contexts involving a child playing with his ball, it means he is throwing it above the ground. In contexts of an ostrich, (*Daḥa*) its home, it means it paved it and removed any pebbles on it. All of this indicates that the meaning of the word *Daḥa* is to remove and pave."

In Zād al-Masīr, ibn al-Jawzī said, "*Daḥa* means to spread." Al-Nasafī said, "*Daḥa* means to spread. The Earth was initially created without spreading. Afterwards, it was spread from Mecca after the heaven had already been created for two thousand years." Al-Baydāwy said, "Allah spread it and paved for inhabitation." Al-Bughawy said, "*Daḥa* is to spread." Ibn 'Attyya said, "*Daḥa* means that Allah spread the Earth." Abu Ḥayān said, "*Daḥa* is to spread." In the commentary of al-Jilālayn, it reads, "*Daha* is to spread but it was not created without being spread before the creation of the heaven." Al-Shawkāny said, "*Daha* is to spread." In his Sufi commentary, ibn 'Ajība said, "*Daha* means to spread and pave it for the inhabitation of its dwellers and their traveling across it. It was spherical, unpaved, when it was first created. The paving started from underneath Mecca two thousand years after the creation of the heaven. The meaning of *Daha* is explained in the following verse, '**He extracted from it its water and its pasture**' [Al-Nāzi'āt:31]. He made the Earth burst with springs, rivers, and pastures where cattle seek their food."

Al-Ṭabrasy, the Shi'i, said, "Allah spread [the Earth] after the creation of the heaven." He also said, "Allah spread and extended it after He built the heaven, raised its ceiling, and proportioned it, and darkened its night and extracted its brightness."

As you have already noticed, most of the commentators maintained that the meaning is to spread. So, what is the source of the "oval" interpretation the contemporaries provided? I think---and Allah knows best---the misunderstanding occurred when they mistook the Arabic word "*udhiya*" of an ostrich for the word "*Daha*," as the former refers to the location where the ostrich lays its eggs—called the dumb nest. They confused the location where the ostrich lays its eggs to the eggs. The statements of the preceding commentators are significantly sounder than those proposed by the contemporaries who are influenced by the theory of the spherical Earth. On a relevant note, the meaning of *Daha* means made it spread like the dumb nest with its high edges made by the ostrich to protect its eggs against the movement of the wind and other reasons for protection. The edges are then raised while the surface where the eggs lay are relatively lower. Does this explanation line up with the Quranic interpretation? I think it does. I do even think that the Earth was first created in the shape of a rubber ball, then Allah spread it by means of His hands, foot, power, or any of His creatures. The Earth was then spread by such stroke, extended and developed high borders. You can try this at home. Make a ball and then pressure it from the middle or strike it with your fist. What happens? The ball will stretch and gush out the insides. If you understand this, you will easily be able to connect the Quranic facts with the Torah, the so-called Bible, the hadiths attributed to the messenger of Allah (peace and blessings be upon him) his companions, the successors, and even the religious myths.

Allah, Almighty, said, **"And the earth—We spread it out and cast therein firmly set mountains and made grow therein [something] of every beautiful kind"** [Qāf:7].

Some contemporary scholars cited this verse as a proof that the Earth is spherical, because it will look spread whenever you look at from any place on Earth. This, however, would not make sense if the Earth has edges, thus they cited the preceding verse as a proof that the Earth is spherical. Nevertheless, this deduction is incorrect, because it is based on a false premise. The first assumption is that the spreading mentioned in the previous verse does not entail infinite spreading. None of the language lexicons explain spreading in infinite terms. Quran, on the other hand, states the opposite and that the spreading is finite. Allah, Exalted and Glorified, said about what will happen at the end of time, **"When the sky has split [open]. And has responded to its Lord and was obligated [to do so]. And when the earth has been extended"** [Al-'Inshiqāq:1–3]. Based on this, the Earth will spread on the Day of Judgment more than it is spread today. Some shortsighted people may argue that there is an opposition between this and the saying of Allah, Exalted and Glorified, **"Have they not seen that We set upon the land, reducing it from its borders? And Allah decides; there is no adjuster of His decision. And He is swift in account"** [Al-Ra'd:41].

My reply is that there is no opposition here, because the verses about spreading, which they think are in opposition to the reduction of the Earth's borders, address the past and not the present. The proof is the saying of Allah, Exalted and Glorified, **"And it is He who spread the earth and placed therein firmly set mountains and rivers; and from all of the fruits He made therein two mates; He causes the night to cover the day. Indeed in that are signs for a people who give though"t** [Al-Ra'd:3]. Allah mentioned He spread the Earth and set the firm mountains and rivers. Do you see this is happening today? Definitely not. The events the verse highlights took place at the beginning of the Earth and what is above it. The same is found in the verse of chapter Qāf where Allah said, **"And the earth—We spread it out**

and cast therein firmly set mountains and made grow therein [some-thing] of every beautiful kind" [Qāf:7]. These verses refer to events that took place in the past. It has no proof that the Earth is spreading now, thereby reconciling any apparent opposition imagined by some people.

Also, the human being abilities are limited. Whenever he is on Earth, it will be spread before him, despite the fact the Earth has borders exactly like the heavens. However, the human being is incapable of perceiving those borders. The problem facing the adoptees of the sphericalness of the Earth is that they perceive the Earth to be very small. In my view, on the other hand, it is massively big, and its mass is equivalent to that of the heavens. If they would realize this, they would not have any problem understanding any Quranic verse nor would they have any of the questions that prove their inability to perceive the due size of the Earth.

Allah, Exalted and Glorified, said, **"[He is] Lord of the two sunrises and Lord of the two sunsets."**

Some people cited this verse to prove the Earth is spherical. They maintain, "Were the Earth to be flat, there would be a single sunset and one sunrise. It would be impossible to have two sunrises and two sunsets or even more." I say, "How ridiculous is this deduction! How bad is the rampant ignorance in our midst?"

Firstly, the reason they maintain such nonsense is that they perceive the sun to be immensely large and a hundred times greater than the size of the Earth, not to mention being hundred miles distant from us. Who said this is factual and taken for granted? They do not have a single proof that this perception is correct but merely theories and thoughts that assumed the appearance of facts.

Dear reader, if your mind grasps the concept that the sun is significantly smaller than the Earth and it is an object that orbits the Earth, you will easily understand the verse. If the sun is smaller, it is not necessary that its sunrise covers the entire Earth the instant it rises. For example, take a huge surface in a dark place, then use the flashlight on your mobile phone or any other source of light. Draw your phone closer to the huge surface and then move it in a circular motion. What will you discover? You will discover that a portion of the surface receives the light while the other portion is still dark. This way, you did not have to move the surface to have a dark and lighted sides. The commentators maintaining the Earth is flat had no problem understanding the verse. They understood many things, none of which entail the Earth is spherical. For example, they maintained that the two sunrises refer to the sunrise in the summer and another in the winter. Others said it refers to the sunrise and the moonrise, the sunset and the moonset. Furthermore, an opinion maintains that it refers to the sunrise, its sunset, dawn, and the dusk. All of these statements are in agreement with the flatness of the Earth, but not necessarily the sphericalness of the Earth.

One rhetorical remark: when Allah addresses His Oneness, He used the sunrise and the sunset in a singular form and the addressee was the Messenger (peace and blessings be upon him). Allah, Exalted and Glorified, said, **"[He is] the Lord of the East and the West; there is no deity except Him, so take Him as Disposer of [your] affairs"** [Al-Muzammil:9]. In Chapter al-Raḥmān, Allah said, **"[He is] Lord of the two sunrises and Lord of the two sunsets."** The addressees in this chapter are two species: humans and jinn, or males and females. Most of the Quranic verses include the mention of pairs, **"The sun and the moon [move] by precise calculation"** [Al-Raḥmān:5]; and, **"He released the two seas, meeting [side by side]; Between them is a barrier [so] neither of them transgresses"** [Al-Raḥmān:19–20]; and, **"From both of them emerge pearl and coral"** [Al-Raḥmān:22]; and, **"Untouched before them by man or jinni"** [Al-Raḥmān:74]. What is even better is that the chapter begins with the name "the Merciful," which is amazingly a single verse by itself. This remark has many great secrets that are difficult for me to elaborate on few pages. If I were to do such an elaboration, I would veer off the purpose of this book. When Allah addressed the group, he said, **"Does every person among them aspire to enter a garden of pleasure? No! Indeed, We have created them from that which they know"** [Al-Maʿārij:38–39]. Allah said afterwards, **"So I swear not by the Lord of [all] risings and settings that indeed We are able to replace them with better than them; and We are not to be outdone"** [Al-Maʿārij:40–41]

In addition to the statements made by the commentators regarding the interpretation of the two sunrises or risings, I have another probable perspective different from the one held by the early and late commentators. What if there are material risings for the sun or portals? At each day, the sun rises from a specific portal, window, or a specific location and sets in another portal or specific location. However, we cannot see those actual portals. Following this interpretation, the sunset and the sunrise represent the very action sun rising and setting, and the two sunrises and two sunsets represent the two main portals, whether you consider them the portals of summer and winter, and the risings and settings represent the various portals. This interpretation is the one I feel most comfortable with. There are material risings and settings, not merely a relativistic issue.

Particularly speaking, Allah, Exalted and Glorified, uses the word "Lord" before every mention of the sunrise and the sunset. Anywhere in the Quran you will find that any subjected entity preceded by the word "Lord" has a material existence. For example, **"the Lord of the Worlds"** refer to all that exists; **"Lord of the heavens and the earth"** [Al-Ra'd:16], which has a physical existence; **"Lord of the Throne"** [Al-Tawbah:129], which is known to most of the Muslims; **"Lord of this House"** [Quraysh:3], which has a physical existence. So, when we find the saying of Allah, Exalted and Glorified, **"Lord of the two sunrises and two sunsets,"** and, **"So I swear not by the Lord of [all] risings and settings,"** it should mean there is a substance. In other words, it has a physical existence. This investigation of the Quranic verses is how I drew this conclusion that the risings and settings are probably material entities like portals or anything alike, and Allah knows best.

While I was researching after I wrote the preceding text, I found a narration attributed to ibn 'Abbās that significantly supports my view, which made me very happy. In fact, every time I read the Quran deliberately and independently, I reach to a conclusion already proposed by ibn 'Abbās. Perhaps this is the reason he was nicknamed "the Interpreter of the Quran." After laboriously reflecting on the Quranic verses and connecting them with one another, I eventually found a similar statement he offered. Abu al-Sheikh related from ibn 'Abbās who said, "The sun has three hundred and sixty apertures through which the sun emerges from a different one on a daily basis, without returning to the same aperture in the next year. It emerges with dislike, saying: 'My Lord, do not make me appear before your servants, as I see them disobey you.'"

A similar statement is made by Abu 'Abzā when commenting on the verse, **"[He is] Lord of the two sunrises and Lord of the two sunsets"** [Al-Raḥmān:17]. He said, "The sun has three hundred and sixty towers on the east and another three hundred and sixty on the west; and it rises and sets from a different one on each day." There is another narration related by Abu al-Sheikh from ibn Ādam, "The sun stays in each tower a month. Each tower is thirty flights of stairs, and between each flight of stairs there is a ritual that decreases until the hour is completed in thirty days. Next, it moves to a next tower."

If they argue, we cannot look at the sun from the west side. If it was rising, we must have seen it. However, they still do not understand that the size of the sun is smaller than what is dictated by the space stations. They also do not understand the limitation of the human vision and the impediments of sight. A human being faces a difficulty when looking at a ship disappearing after a few miles on the horizon, so how could he see the sun from the other side? At any rate, for more information, read the chapter dedicated to the sun.

Allah, Almighty, said, **"Until, when he comes to Us [at Judgement], he says [to his companion], 'Oh, I wish there was between me and you the distance between the two sunrises—how wretched a companion'"** [Al-Zukhruf:38].

Some contemporaries, like Mustafa Maḥmoūd (may Allah have mercy upon him), deduced that this verse prove the Earth to be spherical since the distance on Earth cannot be at its maximum between two sunrises unless the Earth is spherical. Before reconciling between the Quran and modern science, however, there should be a determination of the meaning of the two easts first. This way, the verse will be very understandable if we read one of the early commentaries of those who understood Arabic language well.

Al-Tabary said, "One of those companions told his other companion that he wishes to be distant from him as much as the distance between the two sunrises, namely, the sunrise and the sunset, but the word sunrise included both.

It is said that the two sunrises mean the sunrise of the winter and the sunrise of the summer, because the sun rises from one sunrise in the winter and rises from another one in the summer; so its sunset. It sets at two different sunsets, as Allah, Exalted and Glorified, said, '[**He is] Lord of the two sunrises and Lord of the two sunsets.**' The reason companions are telling this to each other is because they have remained together for so long until they each caused the other to be admitted to hellfire. Those who maintained this opinion narrated that ibn 'Abd al-A'lā narrated from ibn Thawr from Ma'mar from Sa'īd al-Jarīry who said, 'I was informed that when the disbeliever is resurrected on the Day of Judgement from his grave, he will slap Satan with his hands. They would not leave each other's side until they are both admitted to hellfire. At this moment, the disbeliever will say: I wish there is between me and you the distance between the two sunrises; how wretched a companion! In contrast, the believer will have an angel to keep him company until judgment is executed or go wherever Allah wills.'"

Al-Zamakhshary noted, "It means the sunrise and the sunset but the former's name is applied to both. The distance here refers to the space between them. '**That you are [all] sharing in the punishment**' [Al-Zukhruf:39]. This means that sharing the punishment will not avail you as much as those sharing the burden of a difficult matter because of their cooperation and division of labor. The reason being each one of you will suffer punishment beyond his capacity. It is also plausible that this statement is meant to imply a wish of distance between the two companions, at which case the following verse means that such wish will not avail them."

Al-Rāzī said, "They differed over the interpretation of '**the distance between the two sunrises**' into various statements. The first is maintained by most commentators that the distance is between the sunrise and the sunset. It is a linguistic custom for Arabs to name two parallel entities with either name for both. For example, al-Farazdaq composed, 'for us are its moons and the rising stars.' Moons here refer to the sun and the moon. Another example is the name given to both Kufa and Basra 'the two Basras,' the name given to the dawn and afternoon time 'the two 'Asr,' the name given to Abu Bakr and 'Umar 'the two 'Umars,' and the name given to water and dates 'al-'Aswadān.'

"Secondly, astronomers hold that the motion occurring from the sunrise to the sunset is the motion of the Earth while the motion occurring from the sunset to the sunrise is the motion of the still planets and any other mobile celestial body other than the moon. If this is the case, the sunset and the sunrise are each a sunrise in relation to something else. Therefore, using the word 'sunrise' for both meanings is factual.

"Thirdly, this is related to the sunrise of the summer and that of the winter with a great distance between the two sunrises. However, I think this opinion is improbable because the intended meaning of the Allah's saying, '**the distance between the two sunrises**' is to exaggerate the distance, which can only be accomplished when introducing the farthest distance possible. This is not the case with the distance between the sunrise of the summer and that of the winter, thereby this opinion is improbable.

"Fourthly the observable fact is that the daily motion of the sun is rising and then setting. As for the moon, it appears at the beginning of the month from the sunset location and then moves steadily towards the sunrise. This proves that the rising motion of the moon starts at the location where the sun sets. If this is the case, the place called sunrise is considered the place where the sun rises though it is where the moon sets. Conversely, the place called sunset is considered the place where the moon rises though it is where the sun sets. This interpretation has justified why both the sunset and the sunrise are called the two sunrises. Perhaps this interpretation is the nearest possible explanation of the word in addition to attending to all possible scenarios; and Allah knows best."

Al-Alūsi, one of the late commentators, said, "'**the distance between the two sunrises**' means the space separating between both of them, namely, the sunrise and the sunset; this opinion is held by al-Zaggāg and al-Farrā' among others. However, the sunrise outdone the sunset, thus it was rendered in the dual form and the distance is set between both sunrises. In contrast, the original meaning asserts the distance between the sunrise and the sunset, but the brief form is chosen to avoid confusion, because it is obvious that being distant from one of them, since any distance away from one of them draws you closer to the other one. Given they are parallel, the distance of the sunrise from the sunset and vice versa is the farthest distance possible. It is true the context gives the impression of exaggeration, hence no confusion there as well."

Ibn 'Āshūr said, "The two sunrises are the sunrise and the sunset, but the former gained dominance because it is more frequent given the human longing to the sunrise after darkness. The meaning of the sunrise and the sunset is either the location of both of them on the horizon or the earthly location that appears to some of the Earth population where the sun appears to rise from and sets at. According to both interpretations, this represents the farthest distance. The original sentence structure is the distance from the sunrise to the sunset and from the sunset to the sunrise, but the word 'sunrises' replaced all of this."

Al-Sha'raawi said, "The distance meant here is the distance from the sunrise to the sunset. this linguistic style is called '*Taghlīb* (i.e. overriding),' used by Arabs to describe parallels as this parallel of sunrise and sunset. A similar example is when they 'the two 'Umars' to refer to Abu Bakr and 'Umar. Interestingly, looking at the sunrise and the sunset from a geographical and astronomical angel will show that the location of sunrise is a location of sunset for some people and the location of sunset is the location of sunrise for other people. Accordingly, there are two sunrises and two sunsets."

Based on the preceding statements, there is no need to reconcile the Quranic verse with the claims of modern science. The verse leans more to the perspective of the flatness of the earth rather than its sphericalness.

Allah, Almighty, said, **"He created the heavens and earth in truth. He wraps the night over the day and wraps the day over the night and has subjected the sun and the moon, each running [its course] for a specified term. Unquestionably, He is the Exalted in Might, the Perpetual Forgiver"** [Al-Zumar:5].

This verse is the most quoted verse used by late scholars to support the sphericalness of the Earth. They intended to prove the scientific inimitability of the Quran. Nevertheless, had they had any background about modern science, the motion of the Earth, and how day and night take place, they would not have used this verse to prove their theory. This may drive some of the Western scholars venerated by some of our brother to mock them as well as the Book of their Lord. God forbid that His Book become an object of ridicule by any believer or any rational person, to say the least. Unfortunately, by virtue of imitation, some rational people and believers have been led to follow the ridiculousness spread by some contemporaries claiming the verse to be indicative of the sphericalness of the Earth. They said the shape moving at night and day is a spherical shape and they blend in one another.

Before explaining the verse, I must ask those late scholars some questions. "What part of the verse indicates the Earth is spherical?" There is no indication in the verse of the sort. Rather, the verse addresses the night and day. It speaks about *Takwīr* (i.e., wrapping) the night and day. Are you familiar with the linguistic meaning of the word *Takwīr?* It means to coil something in a circular manner, such as wrapping a turban around the head. As for the Quranic context, that explains itself by the means of other verses. Allah, Almighty, said, **"He causes the night to enter the day and causes the day to enter the night, and he is Knowing of that within the breasts"** [Al-Ḥadīd:6]. Let me present you some of the relevant statements of the early scholars who did understand Arabic language:

Al-Ṭabary said, "Either one of them envelopes the other, just as mentioned in Allah's saying, '**He causes the night to enter the day and causes the day to enter the night.**' In the same lines, interpreters have said similar statements. 'Ali narrated to me from Abu Ṣāliḥ from Mu'āwiyya from 'Ali from ibn 'Abbās who said, '**He causes the night to enter the day and causes the day to enter the night**' means that He, Almighty, let the night seeps through the day. Muḥammad ibn 'Amr narrated to me from Abu 'Āṣim from 'Īsā and al-Ḥārith from al-Ḥasan from Warqā' from ibn Abu Nujayḥ from Mujāhid who said the meaning of Allah's saying, '**He causes the night to enter the day**' is to let it fall on it.

"Bishr narrated to us from Yazīd from Sa'īd from Qatāda who said the meaning of '**He causes the night to enter the day and causes the day to enter the night**' is to make each one of them envelopes the other. Muḥammad narrated to us from Aḥmad from 'Asbāṭ from al-Suddy who said the meaning of '**He causes the night to enter the day and causes the day to enter the night**' is to bring the day and dispels the night as well as bringing the night and dispelling the day. Yūnus narrated to me from ibn Wahb from ibn Zayd who said the meaning of Allah's saying, '**He causes the night to enter the day and causes the day to enter the night**' is to dispel the night and envelopes it with the day as well as dispelling the day and envelopes it with the night."

Al-Qurtuby said, "Al-Ḍaḥḥāk said it means to let each one of them fall on the other, which is the linguistic meaning of *Takwīr*, particularly to let objects fall on one another. When added to one's luggage, the word *Takwīr* means to put it on top of each other. When added to the turban, it means to roll it up." This meaning is reported to have been explained by ibn 'Abbās regarding the interpretation of this verse. He said, "What is reduced from the night is transformed into the day and whatever is reduced from the day is transformed into the night. This is the meaning of Allah's saying, '**He causes the night to enter the day and causes the day to enter the night**' [Fāṭir:13]. It is said that the night wraps the day until the daylight vanishes and the day wraps the night until it dispels its darkness. This is the opinion of Qatāda and it is the meaning of Allah's, Almighty, saying, '**He covers the night with the day, [another night] chasing it rapidly**'" [Al-A'rāf:54].

Ibn Kathīr said, "Allah subjected them and they follow one another tirelessly. Each of them is chasing the other rapidly, as stated by Allah, Almighty, '**He covers the night with the day, [another night] chasing it rapidly.**' This is the meaning of what ibn 'Abbās (may Allah be pleased with them), Mujāhid, Qatāda, al-Suddy and others narrated."

As for the famous linguistic al-Zamakhshary, he said, "*Takwīr* is to roll up something like a turban on the head. It can be interpreted in various ways: the day and night are successive such that whenever one goes the other follows. Such process is similar to the idea that one envelops the other just as one's cloth envelope the individual. Other interpretation suggests that each one of them dispels the other when it comes. The way each covers one another is likened to something tangible that covers it from being sighted. Moreover, another interpretation holds that each one of them rolls up over the other successively such as the rolling up the folds of a turban on top of each other."

Al-Rāzī said, "The meaning of this *Takwīr* is that any increase in one of them is a decrease for the other. The meaning of wrapping the day and night is mentioned in the hadith, 'We seek refuge in Allah from al-ḥawr (decline) after

Kawr (rectification).' It means we seek refuge from retraction after advancement. Know that Allah, Exalted and Glorified, expressed this meaning in His saying, '**wraps the night over the day;**' and, '**He covers the night with the day, [another night] chasing it rapidly;**' and, '**He causes the night to enter the day and causes the day to enter the night;**' and, '**And it is He who has made the night and the day in succession for whoever desires to remember**'" [Al-Furqān:62].

Al-Baydāwy said, "Each one of them envelops the other as if it rolls over it like the cloth envelops the person. Or the meaning is to be covered with a covering; or the meaning is to roll up over it like rolling the folds of the turban on top of each other."

Al-Fayrūzābādī said, "The night is wrapped around the day such that the day is longer than the night as well as the day is wrapped around the night such that the night is longer than the day." In Baḥr al-'Ulūm, al-Samrqandy said, "The verse means that [Allah] rolls the day over the night. Muqātil said it means to empower one over the other so as each reduces the other. Al-Kalby said that it means to impart the night some of the day so that the night is longer than the day and impart the day some of the night so that the day is longer than the night. Al-Qatby said it makes each enter the other. The origin of *Takwīr* is to roll up like rolling up the turban."

Most of the early commentators provided similar interpretations, except for the late ones (may Allah have mercy upon them) who believed in the sphericalness theory of the Earth and used the preceding verse to support it.

Have you ever seen the Yen and Yang symbol on your screen, with the dark and the white entwined together? All I want you to do is to place the screen on the floor and imagine the white and the dark are moving above a flat surface. This is how they wrap one another, which is the intended meaning of the verse. It lines up with the linguistic meaning, as understood by the Quran commentators and linguists. There is no need to make the Earth spherical, spinning with a velocity of one thousand miles, just to understand the meaning of the verse. Interestingly, the sphericalness theory does not line up with the verse because the day, according to the theory, is one side while the night is on the other without merging in one another. The motion is made by the Earth, thus resulting in day and light.

I would like to draw attention to the following verse. Allah, Exalted and Glorified, said, "**When the sun is wrapped up [in darkness]**" [Al-Takwīr:1]. What is the meaning of this verse in your perspective? Is the sun spherical today? Or will it be spherical on the Day of Judgment? What is the meaning of wrapping in this verse? If your interpretation of wrapping veered off your interpretation of the wrapping of the day and night, what would be the reason for such a change? I will stop here, because I want you to research it yourself. Do not be lazy and feel disabled. If you endeavor to research it yourself, you will discover a lot of new things you were unaware of.

Allah, Almighty, said, "**O company of jinn and mankind, if you are able to pass beyond the regions (diameter) of the heavens and the earth, then pass. You will not pass except by authority [from Allah]**" [Al-Raḥmān:33]

Some commentators cited this verse as a proof for the sphericalness of the Earth, because a diameter can only be found in a circular and spherical shape. Because of their heedlessness, they have not considered that Allah, Exalted and Glorified, said the regions (and/or diameter) of the heavens and Earth, not the diameter of the Earth, which means the total of the Earth and heavens. I mentioned earlier that the total of heavens and Earth can be spherical, like a bubble or a cosmic egg. If you could imagine this, you would be able to understand why it is not possible to penetrate through the heavens and Earth. As for their interpretation of the word "authority" as knowledge, this interpretation is inaccurate for anyone who investigates the Quranic verses. Yet, knowledge can be a part of it as well. Those who claimed they penetrated through the Earth by means of knowledge, they ignored the following verse, "**There will be sent upon you a flame of fire and smoke, and you will not defend yourselves**" [Al-Raḥmān:35]. If they claim that they evaded the flame of fire and smoke by virtue of the authority of knowledge, I say to them, "Prove to us that one of you who failed to penetrate through the regions of the heavens and Earth evaded the flame of fire and smoke. Or is this great authority granted to all creation?"

Allah, Almighty, said, "**And you see the mountains, thinking them rigid, while they will pass as the passing of clouds. [It is] the work of**

Allah, who perfected all things. Indeed, He is Acquainted with that which you do" [Al-Naml:88].

The Quran commentators differed over the meaning of this verse, whether this happens in this world or at the end of time? Whichever may be the correct answer, some of the contemporaries cited this verse as a proof for the sphericalness of the Earth. Their deduction, however, is out of context and incorrect, because Allah speaks about the mountains rather than the Earth. Those believing in the flatness of the Earth thought they are in a pickle. How could the mountains move while the Earth is still? Does this mean the theory of sphericalness is true?" In fact, this fear is one of the causes of the problem. When fear strikes, people tend to avoid the literal meaning of the Quranic verses and hold distant interpretations, as if the individual desires to understand everything and is not ignorant about anything. If one feels he is ignorant of the meaning, he will choose another meaning to cover this ignorance. If we assume they choose a meaning, and an incorrect one, they are boldening their ignorance. They should have just believed in the speech of Allah. Since He said the mountains pass, they pass without reconsidering it as long as one has not learned its actual meaning yet. The interpretation provided by the believers of the sphericalness theory are unconvincing. Until the day comes when there is a convincing interpretation, I will just believe that the mountains pass like clouds. This does not happen often, as it requires a very strong belief in the verses of the Quran.

After researching the commentaries, I think the most likely correct interpretation of this verse is that of Muḥammad al-'Amīn Al-Shanqīṭy, the author of 'Aḍwā' al-Bayān, and al-Ṭāhir ibn ʿĀshūr, the author of al-Taḥrīr wa al-Tanwīr. Al-Shanqīṭy commented on this verse as follows, "In the introduction to this Blessed Book (Quran), I mentioned that one type of its eloquence is when a scholar suggests a meaning of a verse though the verse itself includes a proof invalidating this suggestion. Moreover, another type of its eloquence is determining the meaning in light of the collective number of the relevant verses, which is very common in the Quran, because this commonality shows that the meaning is needless of outside sources to determine it. We have provided man examples from the Quran to prove the two types of eloquence. One of those is this verse in chapter al-Naml. Some people claimed that the saying of Allah,

Almighty, 'And you see the mountains, thinking them rigid, while they will pass as the passing of clouds' prove that the mountains seem rigid and still for who sight it in this world though it passes like clouds. The two types of eloquence disprove this suggestion. When measuring this suggestion against the first type, it shows that the saying of Allah, 'and you see the mountains' is related to the saying of Allah's saying, 'And [warn of] the Day the Horn will be blown, and whoever is in the heavens and whoever is on the earth will be terrified' [Al-Naml:87]. It means that when the Day the Horn is blown comes, those on the heavens will be terrified and the mountains will pass. This clear proof shows that the passing of the mountains like clouds will be on the Day when the Horn is blown.

"As for measuring it against the second type of Quranic eloquence, all the verses addressing the movement of the mountains speak about the Day of Judgement. For example, Allah, Almighty, said, 'On the Day the heaven will sway with circular motion. And the mountains will pass on, departing' [Al-Ṭūr:9–10]; and, 'And [warn of] the Day when We will move the mountains and you will see the earth prominent' [Al-Kahf:47]; and, 'And the mountains are removed and will be [but] a mirage' [Al-Naba':20]; and, 'And when the mountains are moved' [Al-Takwīr:3]. In addition, Allah, Almighty, said on this verse at hand, '[It is] the work of Allah, who perfected all things.' This meaning is emphasized in many other verses such as, 'So blessed is Allah, the best of creators' [Al-Mu'minūn:14]; and, 'You do not see in the creation of the Most Merciful any inconsistency' [Al-Mulk:3]. Moving, setting, and firming the mountains before they move is a perfect work. Allah, Almighty, concludes this verse with, 'Indeed, He is Acquainted with that which you do.'"

Al-Ṭāhir ibn 'Āshūr said, "The majority of the commentators said that the verse addresses an event that will take place on the Day of Judgement. They have connected the saying of Allah 'And you see the mountains, thinking them rigid, while they will pass as the passing of clouds' to the previous verse, 'And [warn of] the Day the Horn will be blown, and whoever is in the heavens and whoever is on the earth will be terrified.' They considered the action of seeing here to be physical sighting and the passing of the clouds is used to express speed. The point of likening the mountains to the passing of the clouds is to emphasize the commonality of the disengagement of particles and their inflation. This is relevant to the saying of Allah, '**And the mountains will be like wool, fluffed up**' [Al-Qāri'a:5].

"They maintained that the addressee in Allah's saying, '**And you see the mountains**' is unspecified to indicate generalization. They considered the meaning of the verse at hand similar to the verse, '**And [warn of] the Day when We will move the mountains.**' According to the verses, these events will take place before the Day of Resurrection, which would be inconsistent with the provided interpretation, because the verses including the destruction of the mountains shows this will happen at the end of this world just before or after the first Blow. They answered that the mountains will be destroyed at this time then move on the Day of Resurrection, following the saying of Allah, Almighty, '**And they ask you about the mountains, so say, 'My Lord will blow them away with a blast''** [Ṭaha:105], until His saying, '**That Day, everyone will follow [the call of] the Caller [with] no deviation therefrom**' [Ṭaha:108]. The Caller here is Isrāfīl. There are multiple interpretations to the meaning of the words 'follow' and 'call.'

"Some Quran commentators said that his happens by the first Blow; so is all the verses mentioning the destruction and blasting the mountains. In this manner, they have disregarded the connection between the saying of Allah, '**And you see the mountains, thinking them rigid,**' with '**And [warn of] the Day the Horn will be blown, and whoever is in the heavens and whoever is on the earth will be terrified.**' This means that they are separate actions and will not take place on the same Day. Both interpretations considered the saying of Allah, Almighty, '**[It is] the work of Allah, who perfected all things,**' to maximize the power of Allah, Almighty such that blowing in the Trumpet and moving the mountains are manifestations of His power. Their interpretation of the word 'work of Allah' implies the generality of creation without specifying it with the meaning of formation or origination, because the perfection here is a mastery of something whereas the destruction requires no perfection.

"Al-Māwardī maintained that it is an example that Allah has given rather than a statement. As far as this given example is concerned, three views were put forward. They are as follows: First, it is a parable for this worldly life. When one looks at it, he thinks it is rigid and stable while it is moving like clouds. This is the view of Sahl ibn 'Abd Allāh al-Tusturī. Second, it is a parable set forth for faith where one thinks it is firmly fixed in one's heart and that their deeds are raised to the heavens. Third, it is a parable set forth for the body when the soul is pulled out of it and the soul takes its way to the throne. Hence, it seems that they deemed the parable a metaphor. However, any adept critic would realize that these three interpretations are all far-fetched and implausible, because if an object is likened to the mountains, such a state of the mountains is not rigid. Therefore, it cannot be the grounds for the comparison, even if the term 'the mountains' is a metaphor for something and the object that the phrase 'the passing of clouds' is used as a metaphor for is neither explicit nor implicit.

"The discussions of the commentators of this very verse fail to provide a cogent explanation that could clearly show why when one looks at the mountains, one thinks they are firmly fixed, and what the rationale for likening their passing to that of the clouds is. Their discussions also do not provide an elaboration of why the verse is concluded with the saying of Allah, Almighty, '**[It is] the work of Allah, who perfected all things**' [Al-Naml:88]. That is why, this verse has a salient locus that is worthy of contemplation and examination. It is mentioned as a parenthetical sentence in between the concise segment and its ensuing elaboration starting from His saying, '**whoever is in the heavens and whoever is on the earth will be terrified**' [Al-Naml:87], up until His saying, '**Whoever comes [at Judgement] with a good deed will have better than it, and they, from the terror of that Day, will be safe**' [Al-Naml:89]. Such a parenthetical declaration is an indication of the perfect work of Allah, Almighty, in the context of commination and punishment.

"He did the same in the other verse that reads, '**Do they not see that We made the night that they may rest therein and the day giving sight?**' [Al-Naml:86]. The earlier verse could also be a joint complement of the verse '**Do they not see that We made the night that they may rest therein and the day giving sight?**' [Al-Naml:86], and the phrase '**And [warn of]**

the Day the Horn will be blown' [Al-Naml:87] is a parenthetical sentence between them, given that the latter segment of the verse hints at the depiction of life after death. However, this is an invitation to the people of knowledge and wisdom to direct their attention to ponder on what this universe has to offer in terms of sublime pieces of wisdom and perfect forms of creation.

"In fact, this is science contained in the Quran to serve as a miracle that scientists can appreciate from a scientific point of view just as scholars of literature appreciated the way in which the Quran was written and arranged, as we have elaborated in the second section of the tenth conclusion. People once thought that the sun orbits the earth, and this gives rise to the alternation of the day and the night. They thus thought the Earth is stable and firmly fixed. Fortunately, some Greek scholars arrived at the conclusion that it is in fact the Earth revolves around the sun every year and on its axis over the length of the day, and this is what lights almost one half of the Earth and leaves the other half dark. This is what we refer to as the alternation of the day and the night. However, this conclusion was vehemently criticized at the time. The grounds for such a conclusion was the premise that the smaller object should move around the bigger one that would guide its movement, and not the other way around. In actuality, this is such a compelling argument. Since the movement has different orbits, it is possible for the smaller object to revolve around the bigger one as per the naked eye and astrological calculations. This theory only came to light in the seventeenth century at the hands of the Italian mathematician Galileo Galilei.

"Among its myriad signs and right after its proof of the formation of light and darkness, the Quran further suggests another proof that Quran commentators did not elaborate on and thus went unheeded. The Quran linked the movement of the Earth with that of the mountains, because the mountains are the raised edges and sides of the Earth. Therefore, the movement of their shade that keeps diminishing to the minimum before noon and keeps increasing in the afternoon, the observation of the movement of such shades in a way similar to that of ants, and the movement of the peaks of these mountains with the sun disc in the morning and in the evening all reveal that the sun is firmly fixed in its place, as per astronomers and meteorologists. For that reason, the educational style in the saying of Allah, Almighty, '**Do they not see that We made the**

night that they may rest therein' [Al-Naml:86] was changed to an address to them **'And you see the mountains'** [Al-Naml:88], and the addressee is the Prophet Muḥammad (Peace and blessings be upon him). This is an address to him to teach him about something he already knows, and that is why it is an address dedicated to him only and does not apply to others as the general declaration in His saying, **'Do they not see that We made the night that they may rest therein and the day giving sight?'** [Al-Naml:86].

"It is also something kept for the scientists and people of knowledge of his community who are contemporaneous with the emergence of this unerring fact. Allah informed the Prophet (Peace and blessings be upon him) about this prodigious secret of the Earth system as He informed Prophet Abraham (Peace be upon him) of how to resurrect the dead. At the time, Allah informed His messenger (Peace and blessings be upon him) about this very secret and entrusted him with it in the Quran, but He did not command the Prophet to convey it to others since people had no interest to know about that, back then. But when science would disclose such discoveries, the people of the Quran would find that in their scripture too, and they would capitalize on it as deafening evidence for their faith. This interpretation of the verse is more apt since it conforms to His saying, **'And you see the mountains'** [Al-Naml:88], which entails that the person sees them as firmly fixed and stable. As per this interpretation, His saying **'thinking them rigid'** means that they are solid which exquisitely suits the nature of mountains because mountains cannot be fluid. His saying **'while they will pass'** means that 'while they move.' His saying **'like the passing of clouds'** means they do move but in a way that cannot be easily recognized at first sight.

"His saying after that **'[It is] the work of Allah, who perfected all things'** means that this is in terms of their familiar structure and not in terms of the dysfunction of that structure because it is not befitting to describe a dysfunctional structure as a perfected creation. However, it can be described as a matter of gargantuan proportions or any other description that better suits the state of affairs in the hereafter that is not possible to conceive or imagine. The phrase **'the passing of clouds'** refers to the way mountains move, namely they move from one side to another. Yet, if someone were to look at them, they would appear stable and firmly fixed in their place just as if someone were to look at the clouds that are everywhere in the horizon, they would think that they are rigid and not moving while in reality they do move from one direction to another and pour their rain on different places but the person who looks at them may not realize that. Hence, the movement of the mountains here are different from blowing it away as in the other verse that says, '**And [warn of] the Day when We will remove the mountains and you will see the earth prominent**' [Al-Kahf:47]. Specifically, this takes place at the time of the destruction of the earth in the end of times.

"The predicate '**the work of Allah**' is in the accusative case because it is an infinitive form that was used to corroborate the purport of the earlier phrase '**while they will pass as the passing of clouds**,' meaning that Allah is the One who created such perfected creation. This is a magnification of this spectacular system where colossal celestial bodies move vast distances while people think that they are rigid, firmly fixed and stable. Although they move while the people are on top of them, they still cannot sense that. The term *jāmida* means rigid and fixed in its place, and this is the view espoused by Ibn 'Abbās. In Al-Kashshāf lexicon, the term *jāmida* means fixed in its place and not moving, i.e. it is metaphorical. It is a frequently used form of the term that applies equally to the real and the created objects."

This marks the end of the commentaries of al-Shinqītī and Ibn 'Ashūr (May Allah have mercy on them both). Personally, I opine that my interpretation of the passing of the clouds conforms to that of al-Ṭāhir Ibn 'Ashūr and of a number of other commentators. In other words, it refers to the movement of the mountains in this worldly life, and this does not rule out the possibility that the mountains can be blown away in the hereafter. The reason why I hold this view is the clear meaning of the statement of Allah, Exalted and Glorified, "**And you see the mountains, thinking them rigid, while they will pass as the passing of clouds. [It is] the work of Allah, who perfected all things. Indeed, He is Acquainted with that which you do**" [Al-Naml:88]. This is a manifestation of His perfected work in this world in order for the people to heed and ponder. As for the hereafter when sundry dreadful things drastically take place as outlined in the Quran, the time for thinking and reasoning will no longer be available, and the deadline to ponder will have been elapsed. At that time, the sun will be put out and its light will fade away, the seas will be set on fire, and the stars will fall down, to name a few examples. These are gargantuan catastrophic events are the harbingers of the destruction of the entire universe. Therefore, His saying **"the work of Allah"** indicates His exquisitely perfected creation and His knowledge of everything. Perhaps this was the reason to conclude the verse with the following sentence, **"He is Acquainted with that which you do."**

However, someone might argue that the context signifies that the movement of the mountains takes place on the Day of Judgment and that this is corroborated by the verse that precedes it that says, **"And [warn of] the Day the Horn will be blown, and whoever is in the heavens and whoever is on the earth will be terrified except whom Allah wills. And all will come to Him humbled"** [Al-Naml:87]. In response to that, I argue that this view might sound valid at first sight, but if we were to take a closer, deeper look, we will find something else. Let us look at the few verses that directly precede these verses. We will find the following:

Allah, Almighty, said, **"And when the word befalls them, We will bring forth for them a creature from the earth speaking to them, [saying] that the people were, of Our verses, not certain [in faith]. And [warn of] the Day when We will gather from every nation a company of those who deny Our signs, and they will be [driven] in rows. Until, when they arrive [at the place of Judgement], He will say, 'Did you deny My signs while you encompassed them not in knowledge, or what [was it that] you were doing?' And the decree will befall them for the wrong they did, and they will not [be able to] speak. Do they not see that We made the night that they may rest therein and the day giving sight? Indeed in that are signs for a people who believe"** [Al-Naml:82–86].

As you notice, the verses begin by listing some of the Last Day events and conclude with, **"Do they not see that We made the night that they may rest therein and the day giving sight? Indeed in that are signs for a people who believe."** The context clearly informs that day and night are worldly and represent some of Allah's signs. However, they are preceded by events related to the hereafter, thus signifying a connection between them and the verse including the mention of day and night. Likewise, there is one more event related to the hereafter mentioned afterwards, followed by the mention of something worldly, namely the passing of mountains like clouds. The study of the text itself is as crucial as the study of context, not just limited to the preceding and the following verses.

Now, how do mountains pass while we see them rigid? Allah knows how, since He is the One Who established they are passing like clouds, which is sufficient. Assuming we see them pass, He wouldn't have said "you think they are rigid." This means their appearance as rigid is perfectly natural, but Allah draws our attention to something hidden, namely the movement of the mountains. What is the nature of this movement? I do not know exactly, but we cannot certainly say—like some contemporaries maintain—they just pass because it is the Earth that moves and orbits around itself. It is incorrect to say that the sky, mountains, stars, or the trees move by means of the movement of the Earth. According to the perspective of the scientific society, it is the Earth that orbits. In contrast, the literal meaning of the Quranic verses indicates the stillness of the Earth, so how should we act regarding this verse, despite our sight of rigid mountains?

The answer—and Allah knows best—is that it is possible the Earth is placed above the water, which I call the cosmic water. When it was first created, it was shaking like a wooden piece floating on a water surface. It floats, but its movement is shaky by virtue of the movement of the water. After setting weights on top of it, however, it is firmed. Likewise, I think it is likely that Allah, Exalted and Glorified, set the heavy mountains to provide the Earth with firmness. Since the Earth and mountains are above water, the roots of those mountains serving as anchors of ships are the ones moving, with the water being deeper than the roots of mountains. In this manner, the mountains are not rigid above water but rather pass alongside the water. Meanwhile, the Earth above the water is rigid and cannot move by itself. Just as the wind forces the clouds to pass slowly above us, the water serves the same purpose for the mountains such that it forces it to pass like the clouds. Another force may be the wind under the Earth—assuming it exists--and Allah knows best. It is also possible there is another manner of passing that I do not know yet. At any rate, try to comprehend and imagine what I have said. You will see that I have incorporated all the relevant Quranic verses into a single thread and provided you with an answer. In sum, this verse is not a proof of the sphericalness of the Earth. Instead, it is more of a proof of the flatness of the Earth rather than its sphericalness.

Allah, Almighty, said, "**And proclaim to the people the Hajj [pilgrimage]; they will come to you on foot and on every lean camel; they will come from every deep pass**" [Al-Ḥaj:27].

Some contemporaries cited this verse as a proof of the sphericalness of the Earth, using the word "deep." They thought that depth can only be found in a spherical shape. Let us just agree with them for the moment. Assuming Mecca is at the center of the spherical shape, which locations would be considered deep in this spherical shape? Do they mean that people come forth from inside this spherical Earth? Certainty not. They must mean that people come from locations above this Earth, not from inside it. Can they prove where those deep locations are? Their reasoning will shake up. How about the land whose ground level is higher than that of Mecca? Are they included in this verse or not? Does not the modern science maintain that the Earth orbits around itself? Let us then assume that it orbits, and Mecca lies at its bottom. How would the contemporaries' claim line up with the verse? It would not.

Is this contemporary but distant interpretation maintained by any of the Quran commentators? Let us explore what the commentators have said about this part of the verse "deep pass." I should bring to your attention that some of the commentators mentioned below believe the flatness of the Earth while others believe in the possibility of its sphericalness. Despite this, consider their commentary and compare it with the contemporary commentaries who cited this verse as a proof of the sphericalness of the Earth. Ibn ʿAbbās said, "It means they come from a distant location." In another narration, he is reported to have said, "distant." On the same lines, Qatāda said, "they come from a distant location."

Ibn Jarīr al-Tabary said, "They come from every distant location and place." Al-Qurtuby said, "It means a distant location. One relevant sentence is a deep well, which means its bottom is very distant." Ibn Kathīr said, "Deep here means distant, as maintained by Mujāhid, ʿAṭāʾ, al-Suddy, Qatāda, Muqātil ibn Ḥayyān, al-Thawry and others." Al-Zamakhshary said, "Deep means distant." Al-Razī said, "Pass here means a valley between two mountains. It is used widely to mean road in general. The word 'deep' means distant." Al-Fayrūzabādī said, "It means a distant road." In his *al-Nukat wa al-ʿUyūn*, Al-Māwardī said, "Deep means distant." The famous linguist al-Zaggāg said, "Pass means the valley between two mountains."

As for late commentators, al-Shanqity, the author of *Aḍwāʾ al-Bayān*, said, "Pass (*Faj*) means road. Its plural is found in Allah's saying, '**and We made therein [mountain] passes [as] roads that they might be guided**.' Deep means distant. Habitually, deep is used to express distant as far as heading downward such as the phrase a deep well means its bottom is very distant." Ibn ʿĀshūr said, "Pass refers to a valley between two mountains paved for caravans to walk within. Linguistically speaking, it is used more often to refer to a road, because most of the roads leading to Mecca are between mountains. The word 'deep' means distant downwardly. Or it is a simile whereby Mecca is exemplified as a high place and people are rising up to it. This rising applies to any location one is heading towards, just as decline applies to returning from a location. The caravans here are honored by sharing the journey to Mecca with the pilgrims." Even Sheikh Shaʾraawi, who believes in the sphericalness of the Earth, commented on the verse as follows, "They come from every distant road."

As a matter of fact, I believe the meaning of "deep" transcends the meaning of distant, as maintained by al-Shanqity and ibn ʿĀshūr. In my own opinion, I think it is even inaccurate to interpret "deep" as distant. In another instance in the Quran, Allah, Almighty, mentioned the word "distant," which proves there is an extra meaning to the word "deep" that is not included in the word "distant." Allah, Exalted and Glorified, said, "And they had already disbelieved in it before and would assault the unseen from a distant place" [Saba':53]. Allah could have said, "from a distant pass." Instead, Allah used the word "deep." I think—and Allah knows best—that a pass is a valley between two mountains or high places. In this manner, there is a possibility of multiple roads. Because people come riding every lean camel, it becomes necessary that they come from a deep location (facing downward). When climbing up the valley from a low level, there is definitely a difficulty. Thus, there is a connection between this and the claim that the Earth is spherical. It is improper to bend the meaning of the verses to agree with the theory of the sphericalness of the Earth and impart the verse a meaning it does not include.

18

REFUTATION OF SHEIKH MUHAMMAD AL-AMIN AL-SHINQITI OF THE ARGUMENTATION OF SOME CONTEMPORARIES

O NLY A FEW non-famous sheikhs were able to confront the contemporary scholars who interpret the verses of the Quran in a way that conforms to the results of the scientific periodicals. The best who refuted their citations and their claims, in my opinion, is Sheikh Muhammad Al-Amin Al-Shinqiti, author of the exegesis of *Adwa' Al-Bayan*. Therefore, I have dedicated a chapter to quote his speech, which I hope would be beneficial for my dear reader.

With regard to the verse that reads: **"Do you not consider how Allah has created seven heavens in layers. And made the moon therein a [reflected] light and made the sun a burning lamp?"** [Nuh:15–16], the Sheikh wrote in detail about some useful issues. Let us read his speech contemplatively rather than for the purpose of refutation.

The Sheikh said, "It is to be known from these verses that the moon is in the seven layers of the Heaven which Allah has protected from every outcast devil. Thus, there is no doubt that the devils of satellites will come back to the Earth contemptible, debased, after failing to reach to the moon and the Heaven. There is no doubt also that the Heavens in which the moon lies do not mean absolutely whatever is high and above you, even though the term '*Sama*' (sky) or rather ceiling refers linguistically to whatever is high and above you, such as the roof of the house. In this latter sense, Allah, Exalted and Glorified is He, says, 'let him extend a rope to the ceiling' [Al-Hajj:15]. A poet said whatever is high may be called sky. The Heavens are distinguished by the existence of the sun and moon, for Allah has explicitly stated that the moon is in the seven Heavens, because the pronoun (them) in His Saying, 'And made the moon in them a [reflected] light and made the sun a burning lamp?' [Nuh:16], refers to the seven Heavens in layers. It is a linguistic custom used much in the Quran and the speech of the Arab to use a plural form to refer to only some of it. An express evidence of this use is the recitation mode (Qira'at) of Hamzah and Al-Kisa'i with respect to the two uses of the verb 'kill' in His Saying, 'But if they kill you, then kill them' [Al-Baqarah:191]. He who is killed cannot be ordered, after his death, to kill his killer. Thus, the clear meaning of the verse is 'if they kill some of you, then let some others of you kill them,' as is apparent.

"In *Al-Bahr Al-Muhit*, Abu Hayyan said, regarding the interpretation of His Saying, 'And made the moon in them a [reflected] light' [Nuh:16]: 'It is right that the Heavens are the adverb of the place of the moon, because it does not necessitate that the modifier of a place be filled with the modified object, for example, we say,' Zayd is in the city' meaning that he is in a part of it. Let you know that the verse expressly states that the very moon is in the seven layers of Heaven because the term 'made' in the verse means 'rendered' and expresses that the subject and the predicate are the same. For example, when you say, 'I made clay porcelain and made iron a ring,' it is clear that the clay is the porcelain itself, and the iron is the ring itself. Likewise, in His Saying "And made the moon in them a [reflected] light," [Nuh:16], the light made in them is the moon itself. It is not to be understood from the verse, in linguistic terms, that the moon can possibly be out of the seven layers of Heaven and that only the absolute light of the moon is made in it. If that was what is meant, it would have been said, 'And the light of the moon is made in them.' However, His Saying 'And made the moon in them a [reflected] light,' refers expressly that the light made in them is the very moon. It is not permissible to give to the Quran an interpretation other than the intended meaning without evidence. This is clear as Allah, Exalted is He, expressly states in the Quranic Chapter of Al-Furqan that the moon is in the Heaven which has great stars by saying, **'Blessed is He who has placed in the Heaven great stars and placed therein a [burning] lamp and luminous moon'** [Al-Furqan:61]. Allah, Exalted is He, also expressly states in the Quranic Chapter of Al-Hijr that the Heaven with stars in which the moon is placed is the same Heaven which is protected from every outcast devil. **'And We have placed within the heaven great stars and have beautified it for the observers'** [Al-Hijr:16].

"Some people claim that Allah, Exalted is He, referred to the communication between the people of Heaven and Earth in His Saying, **'And of his signs is the creation of the heavens and earth and what He has dispersed throughout them of creatures. And He, for gathering them when He wills, is competent'** [Al-Shura:29]. But the meaning intended is that they will be gathered on the Day of Resurrection as agreed by the exegetes. This is evidenced by His Saying, **'And there is no creature on [or within] the earth or bird that flies with its wings except [that they are] communities like you. We have not neglected in the Register a thing. Then unto their Lord they will be gathered'** [Al-An'am:38].

"That is why the Day of Judgment is called the Day of Assembly in Allah's Saying, **'The Day He will assemble you for the Day of Assembly—that is the Day of Deprivation'** [Al-Taghabun:9]. That is also explained by the plenty of the verses stating that all creatures will be gathered on the Day of Judgment, such as His Saying, **'Indeed in that is a sign for those who fear the punishment of the Hereafter. That is a Day for which the people will be collected, and that is a Day [which will be] witnessed'** [Hud:103]; **'Are to be gathered together for the appointment of a known Day'** [Al-Waqi'ah:50]; **'Allah—there is no deity except Him. He will surely assemble you for [account on] the Day of Resurrection, about which there is no doubt'** [Al-Nisa':87]; **'And [mention] the Day when the heaven will split open with [emerging] clouds, and the angels will be sent down in successive descent'** [Al-Furqan:25]; **'And your Lord has come and the angels, rank upon rank'** [Al-Fajr:22]; and **'and We will gather them and not leave behind from them anyone'** [Al-Kahf:].

"However, some scholars argue that the verse refers to the living creatures that Allah has scattered in the Earth only, and thus the plural pronoun **'throughout them'** is used to refer to only a part of it, namely the Earth. This usage is well known in the Quran and the Arab language. Other scholars said that the creatures in the Heavens are the angels, arguing that creatures include every being that makes a movement. They claimed that the apparent meaning of the holy verse is that Allah has scattered creatures in the Heaven as He did on Earth. There is no doubt that Allah is capable of gathering the people of Heaven and the people of Earth and everything. However, the Quranic verses we have mentioned above clarified that the meaning of gathering them together is collecting and assembling them on the Day of Resurrection and the exegetes have agreed on this interpretation. Have we taken for granted that the verses refer to their gathering in the worldly life, this would not necessitate that the people of the Earth will reach the people of Heaven. It is rationally possible that those in the Heaven descend to those on the Earth because descending is easier than ascending. The claim made by those who have no knowledge of the Book of Allah that His Saying **'O company of jinn and mankind, if you are able to pass beyond the regions of the heavens and the earth, then pass. You will not pass except by authority [from Allah]'** [Al-Rahman:33] refers to the arrival to the heaven. They also allege that **'authority'** in the verse refers to the new knowledge of rockets and satellites. Their claim that the verse may have a proof that they may pass with this knowledge of satellites beyond the regions of the Heavens and the Earth is definitely rejected for many reasons: First, the meaning of the holy verse is that Allah, Exalted and Glorified is He, informs them that there is no escape or refuge from His predestined decision or His will when they are surrounded by the ranks of angels on the Day of Resurrection and whenever they flee to one side they will find the ranks of the angels before them. At that time, they will be told, **'O company of jinn and mankind,'** [Al-An'am:130]. The **authority** referred to in the verse means the argument and proof, or the power and sovereignty, from all of which they will be deprived on the Day of Resurrection. They will have no way out as Allah, Exalted is He, says, **'And your Lord has come and the angels, rank upon** rank,' [Al-Fajr:22]. And, '**And O my people, indeed I fear for you the Day of Calling**' [Ghafir:32].

"Second: Allah had given the Jinn the ability to fly and pass beyond the regions of the Heavens and the Earth, and they used to steal a hearing from the Heaven as Allah, Exalted is He, said about them: '**And we used to sit therein in positions for hearing**' [Al-Jinn:9]. But they were prevented from doing this act after the Prophet, peace be upon him, was sent, as Allah says, '**but whoever listens now will find a burning flame lying in wait for him**' [Al-Jinn:9]. The Jinn were able to reach the Heaven without a need to a rocket or a space ship. Had the meaning of the verse was as those people who had no knowledge of the Book of Allah claimed, Allah would not have said, '**O company of jinn**' because they used to pass beyond the regions of the Heavens and the Earth before the appearance of the alleged authority, i.e. rockets and space ships.

"Third: the aforementioned knowledge that is no more than a man-made industry is too mean to be called an authority by Allah. It does not go beyond the purposes of this worldly life and has nothing to do with the life after death. Moreover, the whole worldly life is not even worth the wing of a mosquito in the sight of Allah. Allah, Exalted is He, declared its meanness in His Saying: '**And if it were not that the people would become one community [of disbelievers], We would have made for those who disbelieve in the Most Merciful—for their houses—ceilings and stairways of silver upon which to mount**' [Az-Zukhruf:33]. Till His Saying '**And the Hereafter with your Lord is for the righteous**' [Az-Zukhruf:35], Allah denied that the knowledge gained by the disbelievers could be real and He confirmed that they know only the outside appearance of the life of the world (i.e. the matters of their livelihood) and said, '**[It is] the promise of Allah. Allah does not fail in His promise, but most of the people do not know**' [Ar-Rum:6]. The skillfulness of some disbelievers in handicrafts is just the same as the skillfulness of some animals in their industry based on Allah's inspiration to them. The bees use hexagonal shape to build their home in a way that perplexes the most skillful engineers. When the engineers wanted to learn from the bees this way of building and kept them in glass structures to see how they build their honeycombs, the bees refused to let them learn their way and painted the glass with honey before beginning their building so that the humans cannot see their way of building as some trustworthy people told us.

"Forth: Had we admitted that the alleged meaning is the true meaning of the verse, we would find that the following verse which says, **'There will be sent upon you a flame of fire and smoke, and you will not defend yourselves'** [Al-Rahman:35] indicates that had they wanted to pass beyond the regions of the Heavens, they would be burnt by this flame of fire and smoke.

"Likewise, the claim made by some of those who have no knowledge of the meaning of the Book of Allah that Allah refers to the communication between the people of the Heavens and the people of the Earth in His Saying **'The Prophet said, 'My Lord knows whatever is said throughout the heaven and earth, and He is the Hearing, the Knowing"** [Al-Anbiya':4] is baseless as the verse was recited with the imperative mode 'Say' in the Quranic recitation of Al-Jumhur and with the past tense **'said, 'My Lord knows"** according to the recitation of Hamzah, Al-Kisa'i, and Hafs from Asim. The Quranic verse does not indicate any such communication neither in terms of conformity, involvement or commitment, because all that the verse indicates is that Allah, Exalted and Glorified is He, orders His Prophet to say that his lord knows all that the people of Heaven and the people of Earth say, according to the recitation of Al-Jumhur and the recitation of Hamzah, Al-Kisa'i, and Hafs. The meaning of the verse is that the Prophet, peace be upon him, reported that his lord, Exalted and Glorified is He, knows all that is said in the Heaven and Earth. This is clear without question. There is no doubt that Allah is All-Knowing of the secrets and public matters of the people of Heaven and Earth: **'Not absent from Him is an atom's weight within the heavens or within the earth or [what is] smaller than that or greater, except that it is in a clear register'** [Saba':3].

"Likewise, those who have no knowledge of the meaning of the Book of Allah claim that Allah indicates that the people of Earth will ascend to the Heavens, one after another, in His Saying, **'[That] you will surely travel from stage to stage,'** [Al-Inshiqaq:19]. Their claim that the meaning of the holy verse is that people will surely travel from one *Tabaq* (Heaven) to another till they ascend above the Heavens is a baseless allegation based on ignorance of the Book of Allah. First, there are two famous recitations for this term; one of them is *Latarkabanna* (travel), according to Ibn Kathir, Hamzah, and Al-Kisa'i. According to this recitation, scholars argue that there are three opinions regarding the subject of the verb *Latarkabanna*: the most famous of which is that the subject is the second person pronoun referring to the Prophet, meaning that 'You, the Prophet of Allah, shall surely experience a *Tabaq* (state) after a state, i.e. you shall raise in ranks,' as *Tabaq* in Arabic language refers to a state.

"Other scholars, including Ibn Mas'ud, Al-Sha'bi, Mujahid, Ibn Abbas in one of his two narrations, and Al-Kalbi, stated that the meaning is: 'You, Muhammad, shall ascend from one heaven to another,' and that took place in the Night Journey. The second scholarly opinion is that the subject is the pronoun referring to the heaven, meaning that 'the Heaven will experience state after state,' i.e. it will sometimes become like oil, murky oil and other times will split open with emerging clouds, and other times will be folded like the folding of a written sheet for the records. The third opinion is that the subject is a pronoun signifying the human referred to in His Saying: **'O mankind, indeed you are laboring toward your Lord with [great] exertion and will meet it'** [Al-Inshiqaq:6]. Thus the meaning is: You the human will experience state after state, from childhood to old age, and from health to illness and vice versa, or from richness to poverty, or vice versa, or from death to life, or from one of the horrors of the Day of Judgment to another, etc. The second recitation according to Nafi', Ibn 'Amir, Abu 'Amr, and 'Asim is *Latarkabunna* and it is a general speech addressing the people mentioned in His Saying, **'Then as for he who is given his record in his right hand,'** [Al-Inshiqaq:7]. And, **'But as for he who is given his record behind his back,'** [Al-Inshiqaq:10]. Thus, the meaning of the verse is that you, people, will experience state after state. You will move in this worldly life from a state to a state and in the Hereafter from one horror to another. It is argued that it is possible in terms of the Arabic language with which the Quran is sent down that the meaning of *Latarkabunna* be: 'You, people, will travel from one Heaven to another until you ascend to the seventh Heaven' as was the interpretation of *Latarkabanna* in respect of the Prophet, peace be upon him. It is argued that if this is possible in the language of the Quran, why shouldn't the verse be interpreted as such? The answer involves three reasons:

"First: The Quran indicates apparently that the meaning of *Tabaq* is the state experienced including death, and the horrors of the Hereafter based on Allah's Saying after that in the following verse, which begins with the resumption particle *fa* (so), which indicate a sequence of events, **'So what is [the matter] with them [that] they do not believe. And when the Quran is recited to them, they do not prostrate [to Allah]?'** [Al-Inshiqaq:20–21]. This is an apparent *Qarinah* (contextual circumstantial evidence) that indicates that the meaning is 'if they experience state after state, and a horror after horror, why shouldn't they believe and get ready for these distresses?' This is supported by the fact that the Arabs call distresses *banatu Tabaq*, as is known in their language.

"Second: The Companions of the Prophet, may Allah be pleased with them, are the first people addressed with the verse and so they are more likely to apply the verse, but no one of them embarked for Heaven according to the consensus of Muslims. This indicated that this is not the true meaning of the verse, for had it been its meaning, the first people addressed with it (i.e. the Companions) would not have been excluded from it without *Qarinah*.

"Third: The aforementioned Quranic verses that explicitly stated that the Heaven is protected from every outcast devil. Thus it becomes clear that the holy verse involves no evidence to indicate that astronauts or aircrafts ascend above the seven Havens and the future will reveal the reality of these lies and invalid allegations.

"Similarly, some of those who have no knowledge of the Book of Allah claim that Allah, Exalted is He, refers to the ascension of the people of the Earth to the Heaven in His Saying: **'And He has subjected to you whatever is in the heavens and whatever is on the earth—all from Him. Indeed in that are signs for a people who give thought'** [Al-Jathiyah:13]. They argue that Allah's subjection of whatever is in the Heavens to the people of the Earth is a proof that they will reach the Heavens. However, the holy verse does not indicate or give evidence to any of their allegations because the Quran explains in so many verses how what in the Heaven is subjected to the people of Earth, pointing out that the sun and moon are subjected for the benefit of humans and for providing light for them so that they may know the number of years and account of time. Allah, Exalted is He, said: '**And He made subject to you the sun and the moon, continuous [in orbit], and subjected to you the night and the day'** [Ibrahim:33]. The benefits of the sun and moon whom Allah has subjected to the people of the Earth can be counted only by Allah, Who said, '**It is He who made the sun a shining light and the moon a derived light and determined for it phases—that you may know the number of years and account [of time]'** [Yunus:5]. He, Exalted is He, also said, '**And We have made the night and day two signs, and We erased the sign of the night and made the sign of the day visible that you may seek bounty from your Lord and may know the number of years and the account [of time]'** [Al-Isra':12]. There are so many verses pointing out this subjection to the people of the Earth. Allah, Exalted is He, also subjected the stars to the people of the Earth so that they can be guided by them through the darkness of the land and sea, as He said, '**and the stars, subjected by His command'** [Al-A'raf:54]. Allah, Exalted is He, also said, '**And it is He who placed for you the stars that you may be guided by them through the darkness of the land and sea'** [Al-An'am:97]. '**And by the stars they are [also] guided'** [Al-Nahl:16]. There are so many verses in the same respect. This is the subjection of what is in the Heaven to the people of the Earth and the best interpretation of the Quran."

Our aforementioned opinion is clarified by the fact that the first people addressed by His Saying: "**And He has subjected to you whatever is in the heavens and whatever is on the earth—all from Him. Indeed in that are signs for a people who give thought**" [Al-Jathiyah:13]—namely the Companions—had only this subjection, pointed out in many verses of the Quran, of what is in the Heavens. Had the alleged subjection of rockets and satellites been meant, it would have clearly applied to the first people addressed by the verse. Likewise, in His Saying, "**And how many a sign within the heavens and earth do they pass over while they, therefrom, are turning away**" [Yusuf:105], their passing over the signs in the Heavens means their looking at them as explained by Allah in many verses, including: "**Do they not look into the realm of the heavens and the earth**" [Al-A'raf:185]; "**Say,**

'Observe what is in the heavens and earth'" [Yunus:101]; and, "**We will show them Our signs in the horizons and within themselves until it becomes clear to them that it is the truth. But is it not sufficient concerning your Lord that He is, over all things, a Witness?**" [Fussilat:53], in addition to many other verses in the same respect.

Having replied to them, he said a nice word: "Let you know—may Allah guide me and you—that playing with the Book of Allah and misinterpreting it in an attempt to correspond with the opinions of the disbelieving Western scholars is not at all in the interest of this worldly life or the Hereafter. It rather entails mischief in both of them. By standing against manipulating and misinterpreting the Book of Allah, we urge all Muslims to exert their utmost efforts in learning the useful worldly sciences while remaining steadfast in their adherence to their religion, as Allah, Exalted is He, says: **'And prepare against them whatever you are able of power'**" [Al-Anfal:60]

In his interpretation of the verse in Al-Furqan, he also said, "We have previously stated that in the Quranic Chapter of Al-Hijr, the Quran explicitly states that the moon is in the constructed heaven and not in the absolute sky which means whatever is high and above you, because Allah has pointed out in Al-Hijr that the Heaven in which He placed great stars is the Heaven protected and the protected Heaven is the one constructed in His Saying: **'And the heaven We constructed with strength, and indeed, We are [its] expander'** [Adh-Dhariyat:47], and in His Saying: **'And constructed above you seven strong [heavens]'** [An-Naba':12]. It is other than the sky which refers to whatever is high and above you. This is clearly explained in Al-Hijr in His Saying: **'And We have placed within the heaven great stars and have beautified it for the observers. And We have protected it from every devil expelled [from the mercy of Allah]'** [Al-Hijr:16–17]. These verses in Al-Hjir are indicative that the Heaven with great stars is the constructed and protected Heaven rather than the sky that is high and above you. If you learned that, you should know that Allah, Exalted is He, has pointed out in this verse in Al-Furqan that the moon is the Heaven in which the stars are placed because Allah says: **'Blessed is He who has placed in the Heaven great stars and placed therein a [burning] lamp and luminous moon'** [Al-Furqan:61].

This is a proof that it is not the absolute space that is high and above you. The Muslim should not turn away from this apparent meaning except for compelling evidence from what Prophet Muhammad has reported. However, no one should quit the apparent meaning of the Glorious Quran except for a convincing well-known proof.

Undoubtedly, those who try to ascend to the moon with their machines and claim that they had landed on its surface, their meanness, weakness, and inability will eventually show when confronting with the might of the Creator of the Heavens and the Earth, Exalted is He. We have previously stated with regard to the Quranic Chapter of Al-Hjir that this is indicated by Allah's Saying: **'Or is theirs the dominion of the heavens and the earth and what is between them? Then let them ascend through [any] ways of access. [They are but] soldiers [who will be] defeated there among the companies [of disbelievers]'** [Sad:10–11]. It would be argued that the verses you gave as evidence that the moon is in the protected Heaven may have probably a well-known Arabic style which entails its being not indicative of what it stated, i.e. the pronoun refers to the term only rather than the meaning. That is to say, His Saying, **'Blessed is He who has placed in the Heaven great stars'** [Al-Furqan:61]. refers to the protected Heaven but the adverb 'therein' in His Saying: **'and placed therein a [burning] lamp and luminous moon'** [Al-Furqan:61]. refers to the absolute meaning of the term sky (*Sama'*) in terms of being linguistically whatever is above you. This is a well-known Arabic structure expressed by the linguists in saying, 'I have a Dirham and a half, i.e. half of another Dirham.' Similarly, Allah's Saying, **'And no aged person is granted [additional] life nor is his lifespan lessened but that it is in a register'** [Fatir:11] means that the lifespan of another person is not to be lessened. Our reply is that this argument is possible but there is no evidence established to support it and the apparent meaning of the Quran should not be forsaken except for compelling evidence and the apparent meaning of the Quran should be more likely followed rather than the opinions and arguments of the disbelievers and their imitators. And Allah knows better."

In conclusion, I say, May Allah reward you well, Sheikh Muhammad Al-Amin Al-Shinqiti, bless your soul, and make you among the people of the

Paradise. You said what was sufficient and your prophecy that the future will reveal the fallacy of their landing on the moon, proved to be true. Their lie became widespread all over the world. The light of truth became bright and was accepted by those who wanted to accept it and was denied by those who did not want to accept it. The Torah also proved that the moon has its own light and is in accordance with Allah's Saying, **"And made the moon in them a [reflected] light and made the sun a burning lamp?"** [Nuh:16].

Dear reader, I will explain some of these points later in the chapter entitled *Bible and Flat earth Theory.*

19

PROPHETIC HADITHS AND

FLAT EARTH THEORY

THERE ARE MANY narrations attributed to the Prophet (peace and blessings of Allah be upon him) proving the Earth is flat. Moreover, the hadiths stated in the books—whether they are *Marfu'ah*[2] or *Mawqufah*[3]—are more than a thousand, which, along with the repeated ones, I counted myself. Also, the number of hadiths that can serve as direct and straightforward proof that the Earth is flat and not spherical is not less than two hundred. Each hadith among them can be discussed at length as I did regarding the Quranic verses I stopped at in this book. That is why I, by Allah's willing, may in the future write a book on hadiths and the Flat earth theory.

Yet, I would like to say: As for the hadiths used as a means to slander the prophetic Sunnah, such as the hadiths of Al-Israa' and Al-Mi'ra, prostration of the sun or the seven earths beneath Allah's Throne, the people swallowed up by the earth, some hadiths that are *Mawqufah, on the beginning of the creation of the heavens, earth and the creation of the sun and the moon, and even the hadiths, that are Da'ifah* (weak), on the iceberg, or Mount Qaf in some reports, all of this

---

2 Hadiths narrated from the Prophet with a connected or disconnected chain of narration. [Translator]

3 Words or deeds narrated from a Companion of the Prophet that are not attributed to the Prophet. [Translator]

becomes reasonable from the Flat earth perspective. Furthermore, the human mind would be impressed by those reports as they seem logical and in conformity the real science. However, a detailed explanation of this issue would take long, and I do not tend to prolong the book. I just want to point out that the issue is deeper and greater than some contemporary intellectuals' opinion that the reports and narrations are mere naivety and backwardness.

20

BIBLE AND FLAT EARTH THEORY

THE FACT THAT the Earth is flat and fixed (or never moves) is clear in the Bible. This explains the debate that occurs between the Church and scientists about the sphericalness and rotation of the Earth. Some may ask, "Why do you quote evidence from the Bible?" However, the question I would ask instead is, "Why should I not do so?"

If you argue by saying, "Because it is distorted!" I then reply, "What is the evidence of the distortion of its letters and words and not its meanings?" All the proofs of its distortion can be refuted, but I am not going to discuss this now, and it is sufficient to state Allah's saying, **"Then bring a scripture from Allah which is more guiding than either of them that I may follow it, if you should be truthful" [Al-qasas:49].** But let us assume that the words and meanings in the Bible are really distorted, so what did drive the distorter to distort what was stated concerning the shape of the Earth even before the emergence of Islam?

If it is said that it is a matter of error and this misconception spread by mistake or misinterpretation, then it had been confirmed in the Bible that the Earth is flat and does not move, and I will then ask why we do not find the correction to this point in the Quran. Has not Allah corrected some misconceptions such as **"Say not: 'Three (trinity)!'" [An-nisāa:171].** There is a biblical misconception regarding the Trinity and deity of the Christ, so Allah stated the truth and made it crystal clear in the Noble Quran. Also, when the misconception that Allah rested after He had created the heavens and earth spread, we find the correction for this misconception in Allah's saying, **"We created the heavens, the earth, and all that is between them in six days without experiencing any fatigue" [Qaf:38].**

Why do not we find such correction concerning the shape of the Earth and heavens and their fixedness in the Quran? Rather, I think you will not find an issue that the Quran entirely agrees with the Bible about other than the issue of the creation of the heavens and the Earth.

Here are other questions. If the Earth is really spherical and moves, should not this be known from the sacred scriptures after the Bible? Why do not we find in the Quran, the Prophetic Sunnah, or the Companions' sayings what indicates that the Earth is spherical or moving? Why do religions differ in many matters, but at the same time are almost unanimous on the fact that the Earth is flat and does not move, and that the sun spins on its axis. Or are all of them unsound, and, thus, they all share the same false facts in this regard?

Unfortunately, the Arabic translations of the Bible are not sufficiently accurate, and those who read the English translations found them more precise. Therefore, you can check the various English translations for the following texts, which will give you a clearer image about the issue. Such translations are available for free on many websites.

Before stating the texts, I expect you have read the story of the Tower of Babel and the people who wanted to build a tower tall enough to reach the heaven. It is a worldly known story, even before it was written down in the Torah or the Old Testament. Anyway, let us begin with Godspeed.

We find in the Book of First Chronicles, Chapter 16, verse 30:

"Fear before him, all the earth: the world also shall be stable, that it be not moved." This is a strong refutation for the theorists of Spherical Earth that rotates. The same applies for the first verse of the Book of Psalms, Chapter 93: *"The LORD reigneth, he is clothed with majesty; the LORD is clothed with strength, wherewith he hath girded himself: the world also is stablished, that it cannot be moved."* Also, there is the tenth verse of the Book of Psalms, Chapter 96: *"Say among the heathen that the LORD reigneth: the world also shall be established that it shall not be moved: he shall judge the people righteously;"* and the fifth verse of the same Book, Chapter 104:

"Who laid the foundations of the earth, that it should not be removed forever." Also, in the Book of Isaiah, Chapter 45, the eighteenth verse, it is stated: *"For thus saith the LORD that created the heavens; God himself that formed the earth and made it; he hath established it, he created it not in vain, he formed it to be inhabited: I am the LORD; and there is none else."* The point here is about the sentence: *"God himself that formed the earth and made it; he hath established it . . . "*

This is in addition to the eleventh verse of the Book of Zechariah, Chapter 1: *"And they answered the angel of the LORD that stood among the myrtle trees, and said, We have walked to and fro through the earth, and, behold, all the earth sitteth still, and is at rest."* Dear reader, do these texts not indicate that the Earth is unmoving and fixed? Or do they indicate that the Earth is spherical rotating in the space? If you argue by saying, "What is meant is that the Earth does not shake!" Then, I will reply, "So, what do you say then about earthquakes?" It is clear that what is meant in these texts is the whole Earth is really stable and fixed as Allah, the Almighty, has stated in the Noble Quran.

How did the creation begin, as stated in the Torah and the Bible, and what is the sequence thereof?

Let us read the following:

"(1) In the beginning God created the heaven and the earth. (2) And the earth was without form, and void; and darkness was upon the face of the deep. And the Spirit of God moved upon the face of the waters. (3) And God said, Let there be light: and there was light. (4) And God saw the light, that it was good: and God divided the light from the darkness. (5) And God called the light Day, and the darkness he called Night. And the evening and the morning were the first day. (6) And God said, Let there be a firmament in the midst of the waters, and let it divide the waters from the waters. (7) And God made the firmament, and divided the waters which were under the firmament from the waters which were above the firmament: and it was so. (8) And God called the firmament Heaven. And the evening and the morning were the second day. (9) And God said, Let the waters under the heaven be gathered together unto one place, and let the dry land appear: and it was so. (10) And God called the dry land Earth; and the gathering together of the waters called the Seas: and God saw that it was good. (11) And God said, Let the earth bring forth grass, the herb yielding seed, and the fruit tree yielding fruit after his kind, whose seed is in itself, upon

the earth: and it was so. (12) And the earth brought forth grass, and herb yielding seed after his kind, and the tree yielding fruit, whose seed was in itself, after his kind: and God saw that it was good. (13) And the evening and the morning were the third day. (14) And God said, Let there be lights in the firmament of the heaven to divide the day from the night; and let them be for signs, and for seasons, and for days, and years: (15) And let them be for lights in the firmament of the heaven to give light upon the earth: and it was so. (16) And God made two great lights; the greater light to rule the day, and the lesser light to rule the night: he made the stars also. (17) And God set them in the firmament of the heaven to give light upon the earth, (18) And to rule over the day and over the night, and to divide the light from the darkness: and God saw that it was good. (19) And the evening and the morning were the fourth day. (20) And God said, Let the waters bring forth abundantly the moving creature that hath life, and fowl that may fly above the earth in the open firmament of heaven. (21) And God created great whales, and every living creature that moveth, which the waters brought forth abundantly, after their kind, and every winged fowl after his kind: and God saw that it was good. (22) And God blessed them, saying, Be fruitful, and multiply, and fill the waters in the seas, and let fowl multiply in the earth. (23) And the evening and the morning were the fifth day. (24) And God said, Let the earth bring forth the living creature after his kind, cattle, and creeping thing, and beast of the earth after his kind: and it was so. (25) And God made the beast of the earth after his kind, and cattle after their kind, and everything that creepeth upon the earth after his kind: and God saw that it was good. (26) And God said, Let us make man in our image, after our likeness: and let them have dominion over the fish of the sea, and over the fowl of the air, and over the cattle, and over all the earth, and over every creeping thing that creepeth upon the earth. (27) So God created man in his own image, in the image of God created he him; male and female created he them. (28) And God blessed them, and God said unto them, Be fruitful, and multiply, and replenish the earth, and subdue it: and have dominion over the fish of the sea, and over the fowl of the air, and over every living thing that moveth upon the earth. (29) And God said, Behold, I have given you every herb bearing seed, which is upon the face of all the earth, and every tree, in the which is the fruit of a tree yielding seed; to you it shall be for meat. (30) And to every beast of the earth, and to every fowl of the air, and to everything that creepeth upon the earth, wherein there is life, I have given every green herb for meat: and it was so. (31) And God saw everything that he had made, and, behold, it was very good. And the evening and the morning were the sixth day."

As you see, the sequence of creation is completely different from and contrasts with the outputs of the contemporary scientific communities; that is to say, Allah created the heavens and the Earth in the beginning, while the scientific community confirms that the heavens were formed before the Earth. This is besides confirming that the Earth was created before the sun and the moon. Ponder on how Allah mentioned that the Earth was without form and stated that it existed before He created the firmament (the heaven). He also confirmed the presence of water above the heaven, and that He then made the dry land appear and brought forth from it pasture and trees. Allah also confirmed the existence of night and day before the creation of the sun and the moon, with no reference to the movement of the Earth, its rotation, or the like. Certainly, the scientific communities do not and will not agree on such a perspective on creation. However, one who reads the sacred texts, contemplates their apparent meanings, and refers to experiments and reason would definitely see that this elaboration on creation and the following sequence of events are logical and reasonable .Yet, does the aforementioned sequence of creation agree with what is stated in the Quran in this regard? Yes, and I am going to mention examples of similarities between the Quranic verses and what is stated in the Torah regarding the issue.

The Quran	The Bible
"All praise is due to Allah, Who created the heavens and the earth..." [Al-anʿām:1]	(1) In the beginning God created the heaven and the earth.
"and made the darkness and the light . . ." [Al-anʿām:1]	(2) And the earth was without form, and void; and darkness was upon the face of the deep... (3) And God said, Let there be light: and there was light. (4) And God saw the light, that it was good: and God divided the light from the darkness.
"And He it is Who has created the heavens and the earth in six Days and His Throne was on the water . . . " [Hud:7]	(2) And the Spirit of God moved upon the face of the waters.

The Quran	The Bible
"Have not those who disbelieve known that the heavens and the earth were joined together as one united piece, then We parted them? And We have made from water every living thing. Will they not then believe?" [Al-anbiyāa:30]	(6) And God said, Let there be a firmament in the midst of the waters, and let it divide the waters from the waters. (7) And God made the firmament, and divided the waters which were under the firmament from the waters which were above the firmament: and it was so. (8) And God called the firmament Heaven.
"And after that He spread the earth. And brought forth therefrom its water and its pasture" [An-nāziʿāt:30–31]	(10) And God called the dry land Earth; and the gathering together of the waters called the Seas: and God saw that it was good. (11) And God said, Let the earth bring forth grass, the herb yielding seed, and the fruit tree yielding fruit after his kind, whose seed is in itself, upon the earth: and it was so. (12) And the earth brought forth grass, and herb yielding seed after his kind, and the tree yielding fruit, whose seed was in itself, after his kind: and God saw that it was good. (13) And the evening and the morning were the third day.
"It is He who made the sun a shining light and the moon a derived light and determined for it phases—that you may know the number of years and account [of time]. Allah has not created this except in truth. He details the signs for a people who know" [Yūnus:5] "And made the moon therein a [reflected] light and made the sun a burning lamp?" [Nūḥ:16]	(14) And God said, Let there be lights in the firmament of the heaven to divide the day from the night; and let them be for signs, and for seasons, and for days, and years: (15) And let them be for lights in the firmament of the heaven to give light upon the earth: and it was so. (16) And God made two great lights; the greater light to rule the day, and the lesser light to rule the night: He made the stars also. (17) And God set them in the firmament of the heaven to give light upon the earth, (18) And to rule over the day and over the night, and to divide the light from the darkness . . .

The Quran	The Bible
"It is Allah who made for you the earth a place of settlement and the sky a ceiling and formed you and perfected your forms and provided you with good things. That is Allah, your Lord; then blessed is Allah, Lord of the worlds" [Ghāfir:64]	(26) And God said, Let us make man in our image, after our likeness: and let them have dominion over the fish of the sea, and over the fowl of the air, and over the cattle, and over all the earth, and over every creeping thing that creepeth upon the earth. (27) So God created man in his own image, in the image of God created he him; male and female created he them. (28) And God blessed them, and God said unto them, Be fruitful, and multiply, and replenish the earth, and subdue it: and have dominion over the fish of the sea, and over the fowl of the air, and over every living thing that moveth upon the earth. (29) And God said, Behold, I have given you every herb bearing seed, which is upon the face of all the earth, and every tree, in the which is the fruit of a tree yielding seed; to you it shall be for meat. (30) And to every beast of the earth, and to every fowl of the air, and to everything that creepeth upon the earth, wherein there is life, I have given every green herb for meat: and it was so. (31) And God saw everything that he had made, and, behold, it was very good. And the evening and the morning were the sixth day."

It is stated in the Book of Genesis, Chapter 2, the first verse: *"Thus the heavens and the earth were finished, and all the host of them,"* which contradicts with the conclusions of modern science that the universe expands and accordingly new stars and galaxies are formed. Moreover, there are current presumptions that new universes are created, but this is contrary to what is mentioned in the Bible that God finished the creation of heavens and earth and all the host of them in six days. But is this in consistency with the Quran? Actually, I do not know. I may be, or there may be an apparent contradiction. Allah, Glorified is He, says in the Noble Quran, **"And the heaven We constructed with strength, and indeed, We are [its] expander" [Ad-dhāriyāt:47]**, yet the majority of exegetes did not state that the expansion here is referred to the heavens. In other Quranic verses, Allah says, **"And [He created] the horses, mules and donkeys for you to ride and [as] adornment. And He creates that which you do not know" [An-naḥ:8]**, and, **"Whoever is within the heavens and earth asks Him; every day He is bringing about a matter" [Ar-raḥmān:29]**. Such verses may refer to creating new creatures within the heavens and the Earth. Yet, we can reconcile between the Torah and the Quran. Thus, we can say that the Arabic word *"Se'ah"* used in the Quranic verse, which translates to "expansion" in the English translation of the verse **"And the heaven We constructed with strength, and indeed, We are [its] expander We are [its] expander"** can refer to "the ability" and that Allah is able to create the heavens and whatever He wills, just as in Allah's saying, where the Arabic word *Musi'* that is derived from the word *"Se'ah"* means capability in the following Quranic verse, **"But give them [a gift of] compensation— the wealthy according to his capability . . .] [Al-baqarah:236]. Also, the Arabic word "wes'aha" that is derived from word** *"Se'ah"* means capacity in the following Quranic verse, **"Allah does not charge a soul except [with that within] its capacity" [Al-baqarah:286]**. This is similar to what is stated in the Book of Job, Chapter 9, the eighth verse: *"Which alone spreadeth out the heavens and treadeth upon the waves of the sea."*

As for the Quranic verses that may be understood to mean new creatures are being created, we can say that there are many creatures today on the Earth that people do not know anything about. And every day and night, a new baby comes to this world that we do not know about. Thus, there is no contradiction between the Biblical text and Quranic text in this regard.

As for the Quranic verse **"every day He is bringing about a matter"** **[Ar-raḥmān:29]**, it may mean, as some exegetes hold, that Allah forgives a sin, relives a hardship, raises the status of some people, lowers the status of others, and the like. In summary, the Quranic verses do support the Biblical texts, while they may also have other meanings. Yet, in all cases, I believe there is no contradiction between the Biblical verses and Quranic verses, but the detailed explanation of this issue will take too long.

The beautiful thing is that the Bible confirms that heaven is a physical object (place) that is solid and extremely hard. Moreover, the more beautiful thing is that the Bible always links the creation of the heavens with God's handiwork. For example, it is stated in the Book of Psalms, Chapter 19, the first verse: "To the chief musician: A Psalm of David." *"The heavens declare the glory of God; and the firmament sheweth his handywork."* It is also stated in the same Book, Chapter 102, verse 25: *"Of old hast thou laid the foundation of the earth: and the heavens are the work of thy hands."*

In the Book of Isaiah, Chapter 45, verse 12, it is stated: *"I have made the earth, and created man upon it: I, even my hands, have stretched out the heavens, and all their host have I commanded."* In the same Book, Chapter 48, verse 13, it is stated: *"Mine hand also hath laid the foundation of the earth, and my right hand hath spanned the heavens: when I call unto them, they stand up together."* There is an interesting text stated in the Book of Job, Chapter 9, verse 8: *"Which alone spreadeth out the heavens, and treadeth upon the waves of the sea."* There is also the eighteenth verse stated in the same Book, Chapter 37, in the New International Version which states: *"Can you join him in spreading out the skies, hard as a mirror of cast bronze?"* It shall be noted that the Arabic translation version of this verse is not adequate.

Do these texts not remind you of the saying of Allah, Glorified is He, in the Noble Quran, **"And the heaven We constructed with strength, and indeed, We are [its] expander"** [Ad-dhāriyāt:47]. Yes, the heaven is a construct, not a mere space, and it is Allah's handiwork, as Allah stated. Also, we find that Allah likened it to a mirror in the Book of Job.

It can be noted that Allah, the Almighty, has mentioned the creation of the heavens and the Earth several times in the Quran. Also, we often find that the word "created" is used when referring to the creation of both of the heavens and the Earth, such as **"All praise is due to Allah, Who created the heavens and the earth..."** [Al-anʿām:1]. But when Allah tells us about the heaven alone, He says, **"And the heaven We constructed with strength, and indeed, We are [its] expander"** [Ad-dhāriyāt:47]; and, **"[He] who made for you**

the earth a bed [spread out] and the sky a ceiling" [Al-Baqarah:22];
and, "And constructed above you seven strong [heavens]" [An-naba:12];
and, "It is Allah who made for you the earth a place of settlement and
the sky a ceiling . . . " [Ghāfir:64]; and also, "Have they not looked at
the heaven above them—how We structured it and adorned it and
[how] it has no rifts?" [Qāf:6]. Allah confirms several times that the heaven
is a construct, yet we insist to hold the view of modern science that the heaven
is a vacuum. I find that particularly odd!

Also, it is stated in the Book of Daniel, Chapter 4, that Nebuchadnezzar
once saw a dream: "(5) I saw a dream which made me afraid, and the thoughts upon my
bed and the visions of my head troubled me. (6) Therefore made I a decree to bring in all
the wise men of Babylon before me, that they might make known unto me the interpretation of the dream. (7) Then came in the magicians, the astrologers, the Chaldeans, and
the soothsayers: and I told the dream before them; but they did not make known unto me
the interpretation thereof. (8) But at the last Daniel came in before me, whose name was
Belteshazzar, according to the name of my God, and in whom is the spirit of the holy gods:
and before him I told the dream, saying, (9) O Belteshazzar, master of the magicians,
because I know that the spirit of the holy gods is in thee, and no secret troubleth thee, tell
me the visions of my dream that I have seen, and the interpretation thereof. (10) Thus were
the visions of mine head in my bed; I saw, and behold a tree in the midst of the earth, and
the height thereof was great. (11) The tree grew, and was strong, and the height thereof
reached unto heaven, and the sight thereof to the end of all the earth."

It is logically known that if the Earth is spherical, then whatever the greatness of the height of the tree is, some side of the tree will not be seen, no matter
how excellent one's sight is! But if the Earth is flat, then the tree can be seen
from all its sides and this will be reasonable. This is similar to the known story
of the attempt of Satan to tempt the Christ, peace be upon him, which is stated
in the Book of Matthew, Chapter 4, verse 8: "Again, the devil taketh him up into an
exceeding high mountain, and sheweth him all the kingdoms of the world, and the glory of
them." There are many interpretations, justifications, and arguments of scientists
in this regard. Yet, what immediately comes to mind when reading this text and
its apparent meaning is that the Earth is flat. This is in addition to what is stated
in the Book of the Revelation to John, Chapter 1, verse 7: "Behold, he cometh with

clouds; and every eye shall see him, and they also which pierced him: and all kindreds of the earth shall wail because of him. Even so, Amen." We also find similar texts indicating the flatness of the Earth in the Prophetic hadiths.

In the Book of Psalms, Chapter 148, verse 4: *"Praise him, ye heavens of heavens, and ye waters that be above the heavens."* This is a proof of the presence of water above the heaven, and not just the water that was above the heaven at the beginning of the creation in Genesis. In Islam, there are also many reports that prove the presence of water above the heaven.

It is stated in the Book of Joshua, Chapter 10, verse 12: *"(12) Then spake Joshua to the LORD in the day when the LORD delivered up the Amorites before the children of Israel, and he said in the sight of Israel, Sun, stand thou still upon Gibeon; and thou, Moon, in the valley of Ajalon. (13) And the sun stood still, and the moon stayed, until the people had avenged themselves upon their enemies. Is not this written in the book of Jasher? So the sun stood still in the midst of heaven, and hasted not to go down about a whole day."* So, if the Earth is the one that moves and revolves around the sun while the sun is fixed, then why do we find these texts that prove the sun is the one that stood still and never state the same concerning the Earth?

It is narrated by Imam Ahmad bin Hanbal in his Musnad that Allah's Messenger (peace and blessings of Allah be upon him) said, "The sun was not held back for any person except Yoosha (peace be upon him) on the day he was marching to Jerusalem." This report was classified as *sahih* (sound) by Ibn Hajar. Also, Shuaib Al Arna'ut said about it, "its *Isnad* (chain of narrators (is *sahih* according to the conditions set by Al-Bukhari." Moreover, the whole story, without mentioning the name of "Yoosha," is stated in Sahih Al-Bukhari.

It is narrated by Abu Huraira: The Prophet, peace and Blessings of Allah be upon him, said, "One of the earlier Prophets carried out a holy military expedition, so he said to his followers, 'Anyone who has married a woman and wants to consummate the marriage, and has not done so yet, should not accompany me; nor should a man who has built a house but has not completed its roof; nor a man who has sheep or pregnant she-camels and is waiting for the birth of their young ones.' So, the prophet carried out the expedition and when he reached that town at the time or nearly at the time of the 'Asr prayer, he said to the sun, 'O sun! You are under Allah's Order and I am under Allah's Order

O Allah! Stop it (i.e. the sun) from setting.' It was stopped till Allah made him victorious. Then he collected the booty and the fire came to burn it, but it did not burn it. He said (to his men), 'Some of you have stolen something from the booty. So one man from every tribe should give me a pledge of allegiance by shaking hands with me.' (They did so and) the hand of a man got stuck over the hand of their prophet. Then that prophet said (to the man), 'The theft has been committed by your people. So all the persons of your tribe should give me the pledge of allegiance by shaking hands with me.' The hands of two or three men got stuck over the hand of their prophet and he said, 'You have committed the theft.' Then they brought a head of gold like the head of a cow and put it there, and the fire came and consumed the booty. The Prophet, peace and blessings of Allah be upon him, added: Then Allah saw our weakness and disability, so He made booty legal for us."

Just as we do not find in the Quran any verse that refers, from near or afar, to the rotation of the Earth in the space, we also will not find any text in the Bible refers to the same matter. Furthermore, both of the Quran and the Bible confirm, in numerous places therein, the fixedness of the Earth and that it does not move except for what is mentioned concerning the earthquakes. As for the movement of the sun and the moon, the size of the stars, and so on, there are several texts in the Bible that all confirm that the sun moves and is smaller than the Earth, and that stars are also smaller than the Earth. The Quran also confirms the movement of the sun.

It is stated in the Book of the Revelation to John, Chapter 6, verses 12–15:

"(12) And I beheld when he had opened the sixth seal, and, lo, there was a great earthquake; and the sun became black as sackcloth of hair, and the moon became as blood; (13) And the stars of heaven fell unto the earth, even as a fig tree casteth her untimely figs, when she is shaken of a mighty wind. (14) And the heaven departed as a scroll when it is rolled together; and every mountain and island were moved out of their places. (15) And the kings of the earth, and the great men, and the rich men, and the chief captains, and the mighty men, and every bondman, and every free man, hid themselves in the dens and in the rocks of the mountains."

The point here is that the stars fell unto the Earth, but the Earth was not destroyed and shattered; and the kings and the rich hid themselves, which indicates that the stars are much smaller than the Earth. Is there any evidence of such events? Yes, there are many known proofs stated in the Noble Quran.

For example, Allah's Saying: **"When the sun is wrapped up [in darkness]; And when the stars fall, dispersing, And when the mountains are removed"** [At: takwīr:1–3].

Ibn Zaid said on the second verse "the stars are overthrown out of the heaven unto the Earth." Ibn Abbas also said, "No star will be left in the heaven on that day, as they all will fall unto the Earth."

What matters is that there are many Islamic, Jewish, and Christian texts stating the events of the fall of the stars and their size, which is much smaller than that of the Earth. They also confirm that whatsoever falls or descends from the heaven settles on the Earth.

Also, there is great similarity between the Bible and the Noble Quran concerning the story of Noah that indicates facts contrary to the stance of the modern scientific communities. For example, it indicates that the heaven has gates. It is stated in King James Bible, in the Book of Genesis, Chapter 7, verse 11: *"In the six hundredth year of Noah's life, in the second month, the seventeenth day of the month, the same day were all the fountains of the great deep broken up, and the windows of heaven were opened,"* apart from what is stated in the same book, Chapter 8, verse 2: *"The fountains also of the deep and the windows of heaven were stopped, and the rain from heaven was restrained."*

In the New English Translation Bible, Genesis (7:11): *"In the six hundredth year of Noah's life, in the second month, on the seventeenth day of the month—on that day all the fountains of the great deep burst open and the floodgates of the heavens were opened."* And, Genesis (8:2): *"The fountains of the deep and the floodgates of heaven were closed, and the rain stopped falling from the sky."*

As you can see, this text is very similar to what is mentioned in the Noble Quran**: "Then We opened the gates of the heaven with rain pouring down. And caused the earth to burst with springs, and the waters met for a matter already predestined"** [Al-Qamar:11–12].

Just ponder on what is stated regarding "the gates" and how the Earth "was burst with springs." If you think that the only time something is mentioned about the gates or windows of heaven is just in the story of Noah, you are wrong.

It is stated about a prophecy in the Book of Isaiah, Chapter 24, verse 18, *"And it shall come to pass, that he who fleeth from the noise of the fear shall fall into the pit; and he that cometh up out of the midst of the pit shall be taken in the snare: for the windows from on high are open, and the foundations of the earth do shake."* The point here is the phrase "the windows from on high."

Just contemplate on the phrase "the foundations of the earth," which is frequently repeated in the Bible. If the Earth is spherical, then what are its foundations? But if the Earth is flat, such words becomes very reasonable and understandable. Also, it is stated in the Book of Psalms, Chapter 75, verse 3: *"The earth and all the inhabitants thereof are dissolved: I bear up the pillars of it. Selah,"* and in the same book, Chapter 104, verse 5: *"Who laid the foundations of the earth, that it should not be removed forever."* This is besides what is stated in the Book of 1st Samuel, Chapter 2, verse 8: *"He raiseth up the poor out of the dust, and lifteth up the beggar from the dunghill, to set them among princes, and to make them inherit the throne of glory: for the pillars of the earth are the LORD's, and he hath set the world upon them,"* the point here is "the pillars of the earth, and he hath set the world upon them." We also find dear reader in the Book of Job, Chapter 9, verse 6: *"Which shaketh the earth out of her place, and the pillars thereof tremble."*

Going back to the point on the gates or windows of heaven, it is stated in the Book of Malachi, Chapter 3, verse 10: *"Bring ye all the tithes into the storehouse, that there may be meat in mine house, and prove me now herewith, saith the LORD of hosts, if I will not open you the windows of heaven, and pour you out a blessing, that there shall not be room enough to receive it,"* the point here is the phrase "the windows of heaven" and in other English translations "the floodgate of heaven." We also find that the gates of heaven are mentioned in the Quran in places other than the story of Noah, such as Allah's Saying: **"And [even] if We opened to them a gate from the heaven and they continued therein to ascend"** [Al-ḥij'r:14], and: **"And the heaven shall be opened, and it will become as gates"** [An-naba:19].

Also, the gates of heaven are mentioned abundantly in the prophetic hadiths; and the well-known story of the Israa and Miraj (the miraculous night journey) is sufficient in this regard.

The point of stating the proofs that the heaven has gates is to confirm that the heaven is a construction and not a mere vacuum and as every construction has a gate, then it is logical and reasonable that the heaven has gates as well. Unfortunately, those who hold the views of the modern scientific community and use these Biblical texts and Quranic verses to attack the Quran and the Bible, unaware that such texts and verses are an indication that their held views that the heaven is a mere vacuum and the Earth is spherical and moves are wrong.

Going back to the point of the story of Babel mentioned in the Torah, we find in the Book of Genesis, Chapter11: *"(1) And the whole earth was of one language, and of one speech. (2) And it came to pass, as they journeyed from the east, that they found a plain in the land of Shinar; and they dwelt there. (3) And they said one to another, Go to, let us make brick, and burn them thoroughly. And they had brick for stone, and slime had they for mortar. (4) And they said, Go to, let us build us a city and a tower, whose top may reach unto heaven; and let us make us a name, lest we be scattered abroad upon the face of the whole earth. (5) And the LORD came down to see the city and the tower, which the children of men builded. (6) And the LORD said, Behold, the people is one, and they have all one language; and this they begin to do: and now nothing will be restrained from them, which they have imagined to do. (7) Go to, let us go down, and there confound their language, that they may not understand one another's speech. (8) So the LORD scattered them abroad from thence upon the face of all the earth: and they left off to build the city. (9) Therefore is the name of it called Babel; because the LORD did there confound the language of all the earth: and from thence did the LORD scatter them abroad upon the face of all the earth."*

There are many proofs that the Earth is flat, and its ceiling is the heavens, such as what is stated in the aforementioned fourth verse. This is apart from what is mentioned regarding the coming down of the Lord, so if the Earth is spherical and the heaven surrounds it, then the direction will be different from a place to a place, that is, the upwardness in the North will be different from that in the South, and the upwardness in the East will be different from that in the West, and so on. However, if the Earth is flat and its ceiling is the heaven, then the point of "the coming down of the Lord or of the angels" will be clearly understandable and reasonable. Furthermore, who studies Judaism will find more other proof in this regard. Also, remember what Pharaoh said, which is stated in the Quran, **"And Pharaoh said, 'O eminent ones, I**

have not known you to have a god other than me. Then ignite for me, O Haman, [a fire] upon the clay and make for me a tower that I may look at the God of Moses. And indeed, I do think he is among the liars'" [Al-Qaṣaṣ:38]. Have you ever wondered where Pharaoh wanted to reach? If the Earth is flat, then what he wanted is reasonable and understandable. It also indicates that the sovereign head of Egypt at that time believed the sky is the ceiling of the Earth, and this was known in Egyptian mythology.

If it is said that Pharaoh was not Egyptian but was of the Hyksos, then this proves there is another party held that the Earth is flat, and the sky is the ceiling thereof.

As for the mentioning of the ends and corners of the Earth, there is quite a number of biblical texts that state this point; and they cannot be interpreted to mean another thing as some Muslims did with the issue of the corners of the Earth mentioned in the Quran. For example, it is stated in the Book of Isaiah, Chapter 11, verse 12: *"And he shall set up an ensign for the nations, and shall assemble the outcasts of Israel, and gather together the dispersed of Judah from the four corners of the earth."* Also, in the same book, Chapter 24, verse 16: *"From the uttermost part of the earth have we heard songs, even glory to the righteous. But I said, My leanness, my leanness, woe unto me! the treacherous dealers have dealt treacherously; yea, the treacherous dealers have dealt very treacherously,"* and in the same book, Chapter 40, verse 28: *"Hast thou not known? hast thou not heard, that the everlasting God, the LORD, the Creator of the ends of the earth, fainteth not, neither is weary? there is no searching of his understanding."* Also, it is stated in the Book of Revelation to John, Chapter 20, verse 8: *"And shall go out to deceive the nations which are in the four quarters of the earth, Gog, and Magog, to gather them together to battle: the number of whom is as the sand of the sea,"* the point here is the phrase "the four quarters of the earth."

We are going to review some texts that "the ends of the Earth" are mentioned in the Book of Deuteronomy, Chapter 28, verse 49: *"The LORD shall bring a nation against thee from far, from the end of the earth, as swift as the eagle flieth; a nation whose tongue thou shalt not understand,"* and in the Book of Psalms, Chapter 48, verse 10: *"According to thy name, O God, so is thy praise unto the ends of the earth: thy right hand is full of righteousness."* In the same book, Chapter 65, verse 5: *"By terrible things in righteousness wilt thou answer us, O God of our salvation; who art the confidence of all the ends of the earth, and of them that are afar off upon the sea.*" Also, it is stated in the Book of Psalms, Chapter 72, verse 8: "*He shall have dominion also from sea to sea, and from the river unto the ends of the earth,*" and in the Book of Isaiah, chapter 45, verse 22: "*Look unto me, and be ye saved, all the ends of the earth: for I am God, and there is none else.*" In addition, it is stated in the Book of Jeremiah, Chapter 51, verse 16: "*When he uttereth his voice, there is a multitude of waters in the heavens; and he causeth the vapours to ascend from the ends of the earth: he maketh lightnings with rain, and bringeth forth the wind out of his treasures.*" The point of attention here is the phrase "the ends of the earth." Also, it is stated in the Book of Revelation to John, Chapter 7, verse 1: "*And after these things I saw four angels standing on the four corners of the earth, holding the four winds of the earth, that the wind should not blow on the earth, nor on the sea, nor on any tree.*"

In the Book of Job, there are several texts that can serve as proofs of the point we are discussing. For example, it is stated in Chapter 28, verse 24: "*For he looketh to the ends of the earth, and seeth under the whole heaven,*" and in chapter 37, verse 3: "*He directeth it under the whole heaven, and his lightning unto the ends of the earth.*" Now, how can you determine the ends of the Earth if the Earth is spherical? Hold a globe and try to point to the ends or corners of the Earth! Surely, all the results will be wrong. However, if the Earth is flat, the point of the ends and corners of the earth will be reasonable and understandable.

It is stated in the Book of Isaiah, Chapter 24, verse 13: "*When thus it shall be in the midst of the land among the people, there shall be as the shaking of an olive tree, and as the gleaning grapes when the vintage is done.*" How can the midst of the Earth be determined if the Earth is spherical and moves a thousand miles per hour? Moreover, how can the upwardness and the downwardness be determined in a heaven that is a mere vacuum surrounding the Earth where there are no directions? It is stated in the Book of Proverbs, Chapter 25, verse 3: "*The heaven for height, and the earth for depth, and the heart of kings is unsearchable,*" and in the Book of Jeremiah, chapter 31, verse 37: "*Thus saith the LORD; If heaven above can be measured, and the foundations of the earth searched out beneath, I will also cast off all the seed of Israel for all that they have done, saith the LORD.*" Here, it shall be noted that since the children of Israel, both the good and the bad, still remain until today and have not disappeared from existence, which means—according to the Bible—the heavens have not yet been measured and the foundations of the Earth have not yet been examined, then these measurements of the Earth's layers, axis, and so forth, which are recorded by some scientists, are just figures based on illusions.

As for the issue of the face of the Earth, there are many Biblical texts that prove the Earth has a face. This is very reasonable if the Earth is flat, while it will be unreasonable if the Erath is spherical. How can the Earth's face—or its back—be determined in such a case? Try to pick up a spherical object, such as a football, a basketball or the like, then tell me where is its face and its back? You will not be able to determine that. Now hold any flat object such as a coin, a disc, or the like, and you will see that you can determine its face and back so easily.

Here are some Biblical texts that mention "the face of the earth," for example the Book of Genesis, Chapter 1, verse 29: *"And God said, Behold, I have given you every herb bearing seed, which is upon the face of all the earth, and every tree, in the which is the fruit of a tree yielding seed; to you it shall be for meat."* In the same book, Chapter 5, verse 1: *"Now it came to pass, when men began to multiply on the face of the earth."* It should be noted that the Arabic translation of this verse is not good, and they did not even translate the word "face," which we find in many English translations of the Bible.

The most important point then is that there are many texts in the Bible that prove the Earth has a face. In the Noble Quran, Allah, Glorified is He, said: **"If Allah were to punish men according to what they deserve. He would not leave on the back of the (earth) a single living creature: but He gives them respite for a stated Term: when their Term expires, verily Allah has in His sight all His Servants"** [Fāṭir:45[. Now, why did Allah say "the back" instead of "the face?" Fakhr al-Din al-Razi answered this in his book, *Mafatih al-Ghayb*, by saying: "if it is argued that how it is said 'face of the earth' and 'back of the earth' for the same side which the creatures are on, although 'the face' is the opposite of 'the back?' We say that the earth is like an animal with heavy loads which are placed on its back. That is why it is said 'on the back of the earth.'"

Also, I want to add a note here, Allah said "the back," not "inside," which indicates that the theory of Hollow Earth, or of the creatures existing in a subterranean realm inside the Earth and living a life like ours, may not be true.

Many contemporary Biblical scholars and their followers invoke a text in the Bible to prove the Earth's sphericalness. It is the verse 22 of the Book of Isaiah, Chapter 40, which states: *"It is he that sitteth upon the circle of the earth, and the inhabitants thereof are as grasshoppers; that stretcheth out the heavens as a curtain, and spreadeth them out as a tent to dwell in."* The Arabic translation of the verse is abysmal, and the "circle of the earth" has been translated into the Arabic word "Kuraa," which in English means "ball." There is a great difference between a circle and a ball. For example, a coin is a circular flat object, while a ball is a spherical shape. So, there is nothing wrong with the Earth being flat in a circular shape, but not in a spherical shape. Certainly, I do not think they would have interpreted this text to have such meaning without being under the pressure of the scientific community and that their holy scripture would contradict the modern science. Therefore, they looked for a way out, which the text itself contradicts with, as it contains "and the inhabitants thereof are as grasshoppers," which indicates that the speaker was viewing from above and thus saw the inhabitants as grasshoppers.

Also, let us ponder on "stretcheth out the heavens as a curtain, and spreadeth them out as a tent to dwell in." The shape of the tent is well known. Moreover, the Hebrew word used for "the heaven" is "vault," the same word used at the beginning of the book of Genesis, which indicates that the speaker is established above it. So, the concept of "the round shape of the Earth" is completely reasonable and that of "roundness of heavens and earth" is logical, and you can refer to what I have stated concerning the cosmic superbubble, which is a matter which is in conformity with the Jewish, Christian, and Islamic texts. Also, none English translations of the Bible have translated the word in this text as "ball" or "sphere," but have used the word "circle," with some using "disc." This is perfectly consistent with the flat earth theory. Moreover, the word "ball" is also used in the same Book, that is, Isaiah, Chapter 22, verse 18, which indicates that the one who wrote the book of Isaiah knew the difference between "ball" and "circle." It is really sad that preachers of the Bible today have ignored many texts that support the flat earth perspective and tried to distort a text in order to please the scientific community. This is similar to what some of the contemporary callers to Islam have done with the Noble Quran.

If you wonder about how the four corners or how its ends can be determined if the Earth is circular and flat, I will say: The Earth is extended and formed to be inhabited, and its center has a round shape. At the same time, the Earth is flat and has corners. Let us test this theory. Take some dough and make a flat shape with it. Place this shape on a hard body, then hit with a fist in the middle of the shape. Now, look at the resultant shape. You will find that the level where you punched is lower than that of the corners. Here is another test. Take a balloon and, before blowing air into it, try to stretch it, making four fixed angles. Then, place your finger in the middle. Now, what is the result? The level of the center of the balloon will be lower than that of the four corners you have fixed. This concept is not widespread among those who hold the flat Earth theory. Just ponder on this deeply, and you will find, God willing, that this is the most consistent with the Quranic verses, Biblical texts, and reality. Besides, it explains many of the natural phenomena.

There are some beautiful texts that serve as a proof in this regard, in the Book of Proverbs, Chapter 8: *"(22) The LORD possessed me in the beginning of his way, before his works of old. (23) I was set up from everlasting, from the beginning, or ever the earth was. (24) When there were no depths, I was brought forth; when there were no fountains abounding with water. (25) Before the mountains were settled, before the hills was I brought forth: (26) While as yet he had not made the earth, nor the fields, nor the highest part of the dust of the world. (27) When he prepared the heavens, I was there: when he set a compass upon the face of the depth: (28) When he established the clouds above: when he strengthened the fountains of the deep: (29) When he gave to the sea his decree, that the waters should not pass his commandment: when he appointed the foundations of the earth: (30) Then I was by him, as one brought up with him: and I was daily his delight, rejoicing always before him."*

These texts strongly challenge of the scientific community and its current views on cosmology.

Finally, and to conclude this chapter, I am going to state for you the complete chapter eighty-two from the Book of Job as it includes great benefits. It proves, from near and afar, that the perspective of the Bible on the Earth is completely different from that of the scientific community. It also confirms how the Bible is consistent with the Quran. Let us contemplate the following verses, and may Allah cause them to lead to the sound views in this regard: *"(1) Then the LORD answered Job out of the whirlwind, and said, (2) Who is this that darkeneth counsel by words without knowledge? (3) Gird up now thy loins like a man; for I will demand of thee, and answer thou me. (4) Where wast thou when I laid the foundations of the earth? declare, if thou hast understanding. (5) Who hath laid the measures thereof, if thou knowest? or who hath stretched the line upon it? (6) Whereupon are the foundations thereof fastened? or who laid the corner stone thereof; (7) When the morning stars sang together, and all the sons of God shouted for joy? (8) Or who shut up the sea with doors, when it brake forth, as if it had issued out of the womb? (9) When I made the cloud the garment thereof, and thick darkness a swaddlingband for it, (10) And brake up for it my decreed place, and set bars and doors, (11) And said, Hitherto shalt thou come, but no further: and here shall thy proud waves be stayed? (12) Hast thou commanded the morning since thy days; and caused the dayspring to know his place; (13) That it might take hold of the ends of the earth, that the wicked might be shaken out of it? (14) It is turned as clay to the seal; and they stand as a garment. (15) And from the wicked their light is withholden, and the high arm shall be broken. (16) Hast thou entered into the springs of the sea? or hast thou walked in the search of the depth? (17) Have the gates of death been opened unto thee? or hast thou seen the doors of the shadow of death? (18) Hast thou perceived the breadth of the earth? declare if thou knowest it all. (19) Where is the way where light dwelleth? and as for darkness, where is the place thereof, (20) That thou shouldest take it to the bound thereof, and that thou shouldest know the paths to the house thereof? (21) Knowest thou it, because thou wast then born? or because the number of thy days is great? (22) Hast thou entered into the treasures of the snow? or hast thou seen the treasures of the hail, (23) Which I have reserved against the time of trouble, against the day of battle and war? (24) By what way is the light parted, which scattereth the east wind upon the earth? (25) Who hath divided a watercourse for the overflowing of waters, or a way for the lightning of thunder; (26) To cause it to rain on the earth, where no man is; on the wilderness, wherein there is no man; (27) To satisfy the desolate and waste ground; and to cause the bud of the tender herb to spring forth? (28) Hath the rain a father? or who hath begotten the drops of dew? (29) Out of whose womb came the ice? and the hoary frost of heaven, who hath gendered it? (30)*

The waters are hid as with a stone, and the face of the deep is frozen. (31) Canst thou bind the sweet influences of Pleiades, or loose the bands of Orion? (32) Canst thou bring forth Mazzaroth in his season? or canst thou guide Arcturus with his sons? (33) Knowest thou the ordinances of heaven? canst thou set the dominion thereof in the earth? (34) Canst thou lift up thy voice to the clouds, that abundance of waters may cover thee? (35) Canst thou send lightnings, that they may go and say unto thee, Here we are? (36) Who hath put wisdom in the inward parts? or who hath given understanding to the heart? (37) Who can number the clouds in wisdom? or who can stay the bottles of heaven, (38) When the dust groweth into hardness, and the clods cleave fast together? (39) Wilt thou hunt the prey for the lion? or fill the appetite of the young lions, (40) When they couch in their dens, and abide in the covert to lie in wait? (41) Who provideth for the raven his food? when his young ones cry unto God, they wander for lack of meat."

The best conclusion of this text is the saying of Allah in the Noble Quran, **"I did not make them witness to the creation of the heavens and the earth or to the creation of themselves, and I would not have taken the misguiders as assistants"** [Al-Kahf:51], and praise is due to Allah Who said about His Book, **"And this is a Book which We have sent down, blessed and confirming what was before it, that you may warn the Mother of Cities and those around it. Those who believe in the Hereafter believe in it, and they are maintaining their prayers"** [Al-An'am:92].

21

RELIGIONS, MYTHOLOGIES AND

THE FLAT EARTH THEORY

I N MANY ANCIENT creeds and mythologies, such as the ancient Egyptians, Mesopotamia, and Babylonians, we find that it was believed the Earth is flat. The same applies to the Sumerian, Chaldean, Assyrian, Greek, Roman, Norse, Celtic, Hindu, Persian, Buddhist, and Jain civilizations. The same view is also held in Judaism, Christianity, and Islam, as well as in various parts of the world, including African, Australian, Chinese, Japanese, Malaysian, Mexican, Mayan, Incan, Viking, Slavic, Roman, and Native American regions.

The textual statements that the Earth is flat, the sky above it is like a dome, or the heaven is a construction, or of the presence of water above the heaven, are many in number that cannot be compiled in a single book. If Allah wills, I will dedicate a book on the ancient mythologies, creeds, and the flat Earth theory, and the textual proofs, in this regard, from the Quran, the Sunna, and the Islamic heritage in general. Until that day comes, I remind you that the purpose of this book is to make you ponder deeply and seek by yourself the reality of the Earth, and if it is spherical or flat.

It will not be beneficial to tell you what the mythologies say in this regard, as they may be of no importance to you, for they are often-doubted ancient beliefs.

22

QUESTIONS AND GENERAL PROOFS

I N THIS CHAPTER, I will state some questions and general proofs that the Earth is flat, or, least not, not spherical. Let us begin, dear reader.

First: Why does the horizon always appear flat to the observer? Why do we see no deviation in it?

Whether you are on the beach, or you look from your home balcony or an airplane window, you always find the Earth is flat and so is the horizon line. Even if you use binoculars and cameras with powerful lenses, you will find the same result. Also, even if you are in a plane at a high altitude, have excellent eyesight, and use modern binoculars, can you see any deviation? Based on experiment, the answer is: You cannot see any deviation. Moreover, if you are in a plane at a high altitude, you would not have to look down to see the flat shape. Just look to the side, and you will find that the horizon is flat. If the Earth is spherical as they say, we would not see such flat shape but because it is flat and expanded, we always see the horizon flat.

Fluids, especially water, always settle flat. What does this mean?

Bring any bottle of water, flip it upside down, tilt it, put it in any position, or shake it. You will find that the surface of the water is flat. This is one of the known physical properties of water that it settles on the flat object. Try this yourself. Bring a ball or a balloon and then pour water over it. You will find that water does not settle on it, but runs down to reach a flat surface to settle flat on it. Have you tried it? If yes, I want you to ask yourself: What makes you believe that such huge amounts of water in this world is stable on a spherical Earth? It shall be noted that water constitutes 70% of the surface, while the land constitutes only 30%. Why do we find seas stable as if settled on a flat surface? For another example, bring a pan you use for cooking, then pour water in it. Now, notice. Where does the water settle, in the top, corners, or a certain side? Yes, you will notice that the water took a flat shape, because this is the property of water. Now, flip the pan upside down. What is the result? The water is poured on the ground, which would have happened if the Earth were spherical and rotating on its axis at a thousand miles per hour.

Since we see with our own eyes that water is flat everywhere in the world, and we do not see, for example, a mountain of water in the middle of an ocean or uneven levels of water therein, we can then conclude that the Earth is flat.

There is an experiment I want you to try yourself. Take your car to a car wash, let one of the workers lift it, then move one of its wheels. Then spray that wheel with water. Will the water remain stable on the wheel or splatter away? Believe me, when you try these simple experiments that are within reach, your conscious will begin to logically analyze the matter.

Let us assume the Earth is spherical, bearing in mind that we all agree on that water always looks for a flat surface to settle on. Now, why do not we find water over mountains? Imagine there is a circle, as if it is the Earth, then draw a triangle on the circle, and the top of the triangle is out of the circle to the right. Then, draw another triangle to the left of the circle and consider these triangles as if they are the mountains on the Earth. If water looks for a flat surface to settle on, and if the Earth is spherical, knowing that there are tons of water in the north of the Earth, if the Earth revolves at a thousand miles per hour, and water must fall down to look for a flat surface to settle on, so water will fall from the right and left of the circle. If they say, this actually happens, I will then say, "This is wrong, because if it were to happen, many countries would disappear off world maps, and if water falls down, we should then find seas above the mountains to the right, north, or even below the Earth!"

If this is justified as due to "gravity," then why are waves and tsunami formed when gravity gets affected by a simple factor? Does the weight of water have no effect? How unreasonable this is! Apply this on the river movement as well.

Also, they say: There are mountains and valleys deep in the sea, so if gravity is the thing that holds water from falling down, then why do we see the water at one level, the depth closest to the center, the water level of which should be different from the other levels, isn't it so??

When you are on an airplane, have you ever wondered why does the pilot not change the movement of the airplane when flying? If the Earth is spherical, why does the pilot not change the altitude of the airplane and keep on the same line? One of my relatives is a pilot, so I do not need anyone to tell me that he has heard that pilots change the altitude as I know for sure that they do not do so. If it is said the gravity is the reason, and it pulls the airplane, I will say, "Why do we not feel so just as we feel any change that occurs while being on the airplane?" If you are so curious and want to know what is the opinion of my relative, the pilot, about the shape of the Earth, he believes the spherical Earth is a lie and the sound view is that the Earth is flat.

WHAT ABOUT THE STARS?

Have you ever asked yourself why the same stars have been above us since we were young until the present moment? If the Earth is spherical, moves at incredible speeds, and rotates around the sun with the rest of the planets, then why do we see today the same stars above us that we used to see when we were children?

Stability of the stellar system has several functions, some of which are stated in the Quran, and Torah and other scriptures. It is said that the Earth revolves around itself and rotates around the sun at high speed, and the sun is in motion at an incredible speed, and so is the galaxy, a matter that makes you suspect those astronomical figures they mention. In the light of all these alleged facts, why do we see Polaris (the North Star) always in its same place? If you say, as some of the ancients said and many follow their opinion, that the Earth is spherical and this is why when a person heads southward, he will see that Polaris will go down little by little until it disappears due to the sphericalness of the Earth. This view is completely illusionary and misleading, as the actual reason is not the sphericalness of the Earth but because the person is simply moving away.

When there is something above you, and you move away from it, even without heading southward, you will find that it goes down gradually until it disappears at the end of the horizon. This is due to the limited extents of our visual acuity, which you can call "the law of perspective." If they use the rotation of the South stars as a proof of their view, I will say that this is just the reflection of the North stars because of the dome of the sky above us. Bring a large spoon that you use to eat or cook, lift it above your head and look at the middle of it. Then lift the right one. Now, what you see? The reflection, is not it? The dome of the sky above us does the same.

In the perspective of the modern science, the Earth rotates almost a thousand miles per hour eastward. If this is the case, why does not the airplane passenger, when anything faces the airplane and interrupts its flying, just wait therein while the airplane stays up in the air until the Earth spins so fast and accordingly he will reach the place he originally intended without the airplane moving?!

Would you, dear reader, follow those who are mind-killed, and thus say, "The atmosphere rotates with the Earth, and we have to go outside the atmosphere." If so, how can you explain the movement and formation of clouds, the flying and movement of the birds, and the falling of rain? All these issues refute the arguments of the unreasonable who hold such illogical views without deep thinking or being aware of the relevant results.

Even if we believe that the Earth and the atmosphere, because of gravity, move at speeds of up to a thousand miles per hour, then the question that arises: "If the airplane or any vehicle flies in the sky at a speed much faster than the speed of movement of the Earth, and if they and the Earth move in the same direction, then how do such vehicles reach their destinations!" I will give you an example. Let us say there are two cars on the same road, heading in the same direction. One of them moves at 100 kilometers per hour and the other moves at 15 kilometers per hour. How will the latter car catch up the former car? Is that possible? Surely, it is not. If you agree on that example, then how can the airplane reach its destination if it moves in the same direction that the Earth moves in? If the airplane flies in the opposite direction of the Earth's rotation, then the time it takes to get to its destination will be greatly reduced because of the Earth's enormous speed. Also, what about the landing? Do pilots calculate the speed of the Earth's movement as they land in order to land in designated places? If the Earth moves at such incredible speed, when a plane flies, its landing zone or sites should then be changed.

Who remembers the extraordinary jump of Felix Baumgartner that dazzled the world and got many Arab imitators more fascinated in the West? I think you do remember it! It took him a number of hours to reach the required altitude. If the Earth really moves at a thousand miles per hour, then a great difference would be expected between Felix's point of jumping and his point of landing. So why did he land in a spot very near to the point of his jumping? It is sad to see some Arab Muslims mocking the Islamic narrations attributed to the Prophet (peace and blessings of Allah be upon him). Why do we not find them making fun of the illogical and unreasonable views held by the imitators of contemporary scientific communities?

It is said the Earth moves at a speed of more than 300 meters per second at the equator. Suppose you are at the equator holding a stone, and you throw it with all of your strength high toward the sky so that when it falls, it will fall on you. Let us also suppose that the falling of stone takes one or two seconds at most. What would happen? Will the stone fall on you or not? Surely you will answer, "It will fall on me." The question I would ask is, "What about the Earth? Is not supposed to move at more than 300 meters per second, and do not the wind and air have speed? Then why should the stone fall on you? What if you pick up the stone that fell on you and throw it West in the opposite direction of the Earth's rotation? How far it will take the stone? Calculate it yourself and then see where the stone fell to realize that the Earth does not rotate, as alleged.

I have always wondered why the East, Middle East, West, South, and North are called so. Why is it said that China, Japan, and India are in the East, and that Italy and France are in the West?

If the Earth is spherical and in a continuous movement, then the directions should be relative. But from ancient times until the present day, whoever lives in the East knows which countries are in the East and whoever lives in the West knows which countries are in the West. So, what is reason? The same applies for the oceans, whose Arabic equivalent is the word "*Moheet*" that literally means "surrounding," Why is "*Moheet*" named so on a spherical earth? This is surely incongruous. But if the Earth is flat and expanded in the middle and seas surround it, then it is a reasonable for "*Moheet*" to be called so.

It is said that the Earth moves at a thousand miles per hour, and if we hold their view that the atmosphere also moves at the same speed and level, and thus no one feels any change occurring, the question that arises: What about distant stars? Do they move at such speed as well? Or does gravity force them to move? If gravity is able to move such stars, so why does gravity not have an effect on the sun, which is nearer to us than those distant stars?

Again, ask yourself: Why do you find birds hovering in the sky as they like? This is besides the fact stated in the hadith, where the Prophet (peace and blessings of Allah be upon him) said about the birds, "They go out hungry in the morning and come back with full bellies in the evening." How long do birds take to find their food or to return to their nests? They take some time in both cases, right? Do birds take more distance returning to their nests than while going out, or do they take the same distance? Does the length of time birds take to return vary from where they go out, due to the Earth's rotation, or is it the same duration? As you also know, it is said the Earth moves at a speed of more than 300 meters per second, so try to do a simple mathematical operation to know that the issue of birds' going out and returning to their nests is greatly problematic for the Earth's rotation theory.

WHAT ABOUT THE WIND?

The *wind* movement is random and not one directional. This is why we use the terms North wind, South wind, East wind, and West wind. The same applies to the clouds. Move a spoon in a cup of coffee, and see what will happen? Imagine there is a movement at a speed of thousand miles per hour, besides the enormous size of the Earth, what should happen in such a case? Then, consider the movement of birds, winds, clouds and even fireworks, toys flown in the air such as kites, or hot air balloons! What do you feel when you are in a street and a big truck passing in front of you? You feel the wind blowing in one direction, right? Why does the wind not move in only one direction but in different directions, and why does the strength of the wind vary, being strong sometimes and weak other times? Because of these variations, winds have several names according to their types. If the Earth rotates in only one direction, this wouldn't have happened, and the wind would have moved in just one direction as well. Think deeply on this point, and forget about the cunning justifications of the other party to answer this question.

This leads me to the point of the clouds' movement. We see clouds sometimes coming from east to west, sometimes from west to east, sometimes from south to north, and sometimes from north to south. How can this happen if the Earth rotates on its axis eastward at a speed of 1,000 mph? Why, then, is the speed of the cloud movement equal on most days unless pushed by any factor such as strong wind? This is difficult for any reasonable person to accept. This is because the reasonable person does not like to force himself to surrender his mind to hypotheses denied by the senses. Not to mention the issue of the clouds' formation, as it is so difficult for any person freeing himself from blind emulation to be convinced of the other party's theories in this regard, in spite of their cunning justifications.

I remember that one day, I was in a discussion with a group of young people, and they were talking about gravity and its great power that prevents tons of seas from flooding, and firmly holds huge buildings, and the like. One of them raised his hand and said, "Gravity has the capacity to hold all these things and prevent me from flying or jumping high, but I can easily move my hand." Have you ever wondered about the secret of this gravity, which can control the movement of all these heavy objects around us but cannot pull insects, birds, planes, and so forth? The same applies to even fish that move freely in the sea, and kites that children fly in the air. So, what is the reason?

Since ancient times, people have traveled or gone to war. As for those who sailed with ships BC and AD, we have not read about any of them thinking of changing the route or taking into account the view that Earth is spherical and acting accordingly. Also, we have not read about those who engaged in wars and used cannons and similar weapons of raising or lowering the cannon muzzles, taking into account the fact that the Earth is moving and spherical. The reason is that the Earth is flat and not spherical. Even in the present time, pilots, sailors and the like, who believe in the theory of the spherical Earth, do not take into account their held view in their trips. The same applies to my relative who is a pilot and many others.

The strange thing is that the sun is about one hundred and fifty million kilometers far from the Earth, the Earth receives its heat from the sun, and the sun is so enormous that it could contain one million and three hundred thousand Earths in it, yet we find that the temperature in Africa is radically different from that in Antarctica! Also, the species of plants and creatures vary from region to region and thus those in the South are different from those in the North, East, and West! Isn't this funny, even though the distance between the two continents—Africa and Antarctica—is not that far! Moreover, when the sun is covered by clouds, we see the beams of its rays scattered in all directions. If the sun is really at this very far distance as stated by the scientific communities, why do we see its beams spreading out or being triangular (Crepuscular Rays) when the sun is behind or above clouds? Furthermore, if the size of the sun is that enormous, then only one column of sunlight is enough to cover the whole earth, so why do we find the columns of sunlight in all directions as if the sun is an object that is much smaller than the Earth?

Try this experiment. Bring an object and make three or four punctures in it, then use your mobile phone's flashlight and notice what will happen when you get the flashlight closer to those punctures? You will find that the rays scatter, diverge, and become triangular. If you take the flashlight away from the punctures, the rays will get closer to each other. This is simple, and anyone can try it. We have seen since we were young until today that the sun rays diverge and do not converge from each other, which indicates the sun is smaller and nearer to us than they claim.

Have you ever asked yourself why you find the sunbeams on the floor and not on the ceiling when the home window is in the direction of the sunrise? Why does this happen even in highland houses? If the Earth revolves around the sun, shouldn't the angle of the sun's rays be different in case the Earth is spherical, rotates on its own axis, and revolves around the sun? In my whole life, I have never seen the sun rays on the ceiling of any room of my house, which indicates the sun is always above us and the Earth is fixed.

Here is another question. If the Earth rotates around its axis and the sun at the same speed every day, then why do we sometimes find that days get longer, and sometimes we find that they get shorter, and the same applies to nights? Why does the timing of the sunrise and sunset differ over the days? If they say, "This difference is because of the rotation of the Earth on its axis," I then have the right to ask, "Why does this difference occur systematically throughout history?"

If you surf the internet, you will find many experiments and images proving that the Earth is flat, and certainly there are also—on the other hand—experiments and images proving that the Earth is spherical. There are scientific and rational explanations of the both parties' experiments, and the probability of truth and lying exists, and so does the possibility of error and delusion. Therefore, it is difficult to rely on such experiments, for if one would say the Earth is flat, another one would say I am certain that it is spherical, so what is the solution?

Science and experiments are a means of getting to the truth, and if the rational and scientific explanations of experiments are proven, experience and academic certificates and the like do not affect the experiments. That is why, it should not be said, "We cannot rely on the experiments proving that the Earth is flat, because those who have carried them out do not work at NASA." Such statements are very similar to that of the narrow-minded who prevent people from contemplating the Quran on the pretext that those people are not world-famous, do not hold relevant academic certificates of exegesis from the Islamic universities, or the conditions of contemplating and understanding the Quran—that may be more than ten—are not applied to them, although Allah has described the Quran as **"clear proofs"** [Al-Baqarah:99], **"[in] a clear Arabic language"** [An-Nahl:103], and **"And We have indeed made the Quran easy to understand and remember"** [Al-Qamar:17]. Monopolizing the truth by a certain group or category is wrong and unfair to truth seekers.

It is a wonder you can see far objects using advanced binoculars. Such a feat would be impossible if the Earth were spherical, because if you consider the ratio of the Earth's curvature to the distance, the far object then should be out of sight, whether with the naked eye or the advanced binoculars. But people today can simply see the distant objects that should not be seen if there is a curvature of Earth as alleged. This can be tried out easily.

Strangely enough, they could not dig in the Earth more than ten miles in spite of their techniques, yet we find them making space rockets and spacecrafts that reach Mars and beyond! What is much stranger is that they say The Earth floats in space and they launch rockets into space? Here, a question arises: "On what factor do these rockets depend on when they move though space? How do they move while there are no winds to push them to go upward or move, and there are no sources of energy to keep them moving on? Besides, chargers or spare batteries will not work out in such a case as there are no electrical charges."

There are many scientific and mathematical problems to consider, once we accept there is a space in the sense of vacuum, and in fact, this hypothesis destroys their theories one after another.

Some intellectuals and scholars invoke the fact that if you travel east, you will reach the starting point of your travel, and the same applies to traveling west. This is possible either on a spherical or flat Earth, as there is no contradiction that the Earth, just as a coin, is flat and has a round surface. Therefore, it is so easy to fly east and return to the starting point, but it is difficult or impossible to fly starting from north and reach the south, or fly starting from south and reach the north, as this contradicts the fact that the Earth is flat.

For this reason, there is no recorded journey to date in the history of mankind, as far as I know, where someone started traveling from the North, headed north, and returned from where he started, or headed south, then headed north, and returned to where he started.

Speaking of gravity, they say that the waves and tides occur because of the gravity of the moon. Here is a question I like to ask. Which has the stronger gravity, the moon or the Earth? If it were the moon, why do not the waves fly to the moon? If the Earth has the stronger gravity because of its size, which is much larger than that of the moon, why do waves move this way? If the movement of waves is linked to the movement of the moon, and the moon has a movement known since ancient times, then why do we find a difference in the waves and that their movement is inconsistent with the moon's movement and almost random? Then why do only the seas get affected by the moon? What about pools and water basins? Why do not they get affected by the moon? Moreover, if the Earth's gravity is stronger than the moon's, then why does the moon move in its orbit, from ancient times to date, without sticking to the Earth?

Is gravity not a force that a great body attracts a small body? We see in movies that once an astronaut is detached from a spacecraft, he floated away into space. The directors forget the gravitational force of the Earth that makes the moon spin on its axis and forget the force of gravity that affects the seas and causes tides. It seems that the gravity of both the Earth and the moon hates the astronauts and thus do not attract them! How strange are those people who believe in scientific concepts in such a blind way?

Also, if you read something of their explanations about the tidal phenomenon, you would be ashamed of your blind belief in their concepts even for one moment, but it is okay to remind people of some of their explanations. You will find that high tides occur in the areas where the face of the Earth meets the face of the moon, because the moon gets closer and water gets attracted to it. We also find that low tides occur at the poles.

You may think about the other side of the Earth and that the tide will be low on it, would it not? It is a logical view, but the answer is no, the tide on the other side of the Earth is not low but high. This is strange, right? Do you know what their explanation of that? They claim the moon attracts the Earth to it a little bit. In other words, the Earth moves away from the water above it, and thus, the tide is high on the other side of the Earth!

Can you imagine that they teach such absurdity to the young in schools! They do not know that this explanation completely destroys their theory, because they in such way negate the presence of the Earth's gravity. Not only that, but they also make the moon's gravity stronger than the Earth's gravity, although the Earth is about fifty times the size of the moon, and also the Earth is more than eighty times more massive than the moon.

Does what they say about the gravity of the moon and its effect on the earth make any sense? It is really so ridiculous and sad, for their explanations are laughable, and it is disappointing how people accept such explanations, believe in them firmly, and thus intellectually fight for them. In fact, if the Earth is spherical and has gravity and if the moon has gravity as well, then the tide on the side of the Earth that does not face the moon should be low. Do you know why? Because the moon is in some direction, and if it has the ability to pull the Earth toward it, then the tide should be low in that side, because the Earth also pulls the water above it, and thus, the water comes between two gravitational forces—the Earth's gravity and the moon's gravity—which both pull the water toward one direction. Do you see how they destroy their theories by their own explanations? Let alone the problems their theories will face if we also add the factor of the sun's gravity, and they will not be able to provide any logical justifications, particularly for the reasonable who deeply analyze and think on everything they read or are told.

WHAT ABOUT THE WAVES?

How are they formed by the moon's gravity? Does the moon's gravity have any effect on the Earth? Certainly, yes. If you say it affects the seas and causes the tide, then the question that should be asked: "Why are there only two tides and not four? Why is the tide not multiplied in the spot facing the moon? The gravity is supposed to be more powerful there, right?" If the Earth is spherical and it and the moon revolve around its axis, and if the moon faces a side or a spot of the Earth and it gets closer to it, then the tide should be multiplied in such a case! There is another point. If the moon causes tides because of its gravity, why do not we find it affecting water ponds, pools, basins, and so on. Why does it only affect the big seas? Or does the moon have a special taste and thus pulls only certain types of water?

Why is there a big difference between the tide in the Pacific Ocean and that in the Atlantic Ocean? In other words, if you are on the isthmus of Panama, you will find that the tide is on one side that is many feet high while the other side has a different unit for measuring. One side is measured in feet and the other in inches. If the moon is the driving factor behind the tides because of its gravity, then what is the explanation for the said phenomenon?

In fact, I have been searching for a long time for an answer of this question from the perspective of the spherical Earth, but there is no proven convincing answer. Everyone has a different answer, most of which are conflicting or make the entire Copernican system turbulent, and so on.

Another question: Have you ever seen the sun's reflection on the sea? Or even the reflection of the moon? How could it be if the Earth is spherical?

The reflection and deviation of the reflection of the sunlight on the water proves that the Earth is flat. On the other hand, if the Earth is spherical and has curvature, then when you look at the sun from a distance, you should not see the reflection in such a case. Moreover, how can the sun's reflection be in a long line as if it is reflected on a flat object?

They say moonlight is only a reflection of the sunlight, and this needs proof because the characteristics and effects of the sunlight differ from the characteristics and effects of the moonlight. If moonlight is just a reflection of sunlight, then it is supposed to have similar characteristics to that of the sun, and I have never heard myself as about a spherical body that reflects light in the way of the moon. However, I have heard about flat or concave objects that do so. Also, the point that the moon is a spherical body should be examined! It is true that it appears to be circular, but that does not prevent the moon from being flat. It may be a rounded disk and not spherical. This is the first point.

The second point is that we often see the moon in the daytime before the sunset, and we can see the sky through it in the sense that we can see the blue color in it, so if it was a reflection of the sun, why would this happen? There are also a lot of experiments where powerful cameras were used at night which prove that you can see the stars through the moon, so how does this happen? There are many experiments in this regard, yet I myself have a different view of the moon that I have stated in the chapter on the moon.

WHAT ABOUT CLOUDS?

There are so many clouds we see flat from the side we observe them; that is, like a flat ceiling for us, so what is the reason? If what they say about that the Earth revolves around itself and about the atmosphere is true, we then shouldn't have seen the flatness of clouds, but we should rather see curvature in them, right? So why do not we see clouds curved, particularly those huge clouds extending several miles?!

WHAT ABOUT LUNAR AND SOLAR ECLIPSES?

I have dedicated a whole chapter on the phenomena of lunar and solar eclipses, but there is no problem in quickly mentioning them, here.

If they say that scientists today know when lunar and solar eclipses would occur, and this is a proof that the Earth is spherical, and that all of their scientific outputs are sound, I then will say, "This is not valid evidence to accept that the Earth is spherical." That is because for thousands of years, even before the spread of the spherical Earth theory, people could predict such matters, although they have believed the Earth is flat and stable. Whoever ponders on history and the outbreak of the spherical Earth theory will find that it began in China. Besides, whoever reads in the Chinese history and how they could predict the phenomena of lunar and solar eclipses will know that the prediction of such phenomena has nothing to do with the spherical Earth theory. Knowing the times of lunar and solar eclipses is due to studying the occurrences of the past eclipses and of the patterns, a matter through which future eclipses can be predicted.

WHAT IS THE SECRET OF THE CONSTELLATIONS (THE ZODIACAL SIGNS)?

Why have people seen them for thousands of years, relying on them in their predictions, etc.? If the Earth is spherical and moves, as do the sun and the galaxy, then why do the constellations stay in their places? Why do people, to the present moment, follow them, and why are horoscopes published in newspapers, magazines, and on media in general?

HOW DO SPACE ROCKETS FLY IN SPACE WHEN THEY COME OUT OF THE ATMOSPHERE?

How can they fly straight without air? Doesn't this raise many questions about the alleged vacuum? And where do these rockets and this effective capability making them keep up flying go? Why do not we find live streaming from the moment of launching them until they reach the space without changing the angles of the cameras? Also, why do space rockets curve off after they are launched upward? Why do not they go up straight without curving off?

Let us suppose technology has developed to a great extent, and we have invented a lot of extraordinary inventions and thus we have become able to dig deep into the Earth and to make a tunnel linking the farthest north to the farthest south. Also, let us suppose that we wore protective suits and jumped into such a tunnel to reach the other side of the Earth, do we reach there while in a right-side up position or in an upside down position, because people on the other side, according to the spherical earth perspective, are supposed to be in an upside down position? Or will something happen, and we will be upside down to be in conformity with the spherical earth theory?

WHY CANNOT WE SEE THE SATELLITES AND SPACE STATIONS?

Why do the images of the space stations seem unclear and like holograms, while there are clear images of Mars? Besides, they recently have claimed that they captured photos of the Black Hole. Why do not the satellites burn up in space? Have you ever wondered, in the first instance, how they work? Why are not the satellite transmitter dishes that detect frequencies directed upward? Or does the direction of the waves in heaven differ from that on earth? I have dedicated a chapter on satellites and the image of the Black Hole, so do check that.

WHY DOES NOT THE DURATION OF THE FLIGHT DIFFER BACK AND FORTH?

For example, if you travel from the UAE to Saudi Arabia, or from the UAE to India back and forth, you will find the flight duration is the same back and forth or has a very minute difference. What is the reason for this? If the Earth is spherical and rotates at a thousand miles per hour, then this should have a big impact on the round-trip flights.

Moreover, we do not feel the movement of the Earth when we are in a car, so what about those who suffer motion sickness and are with us in the car?

Do they feel the movement of the Earth as they are extremely sensitive to any motion? When you ask the spherical Earth theorists what the Earth does in space, and why does not if fall when floating in space, they reply, "The Earth does fall because of gravity." The huge size and mass of the sun pulls the Earth, so we are in a steady fall, but because of the Earth's lateral impetus, we fall toward the sun, but we avoid it. That is, we keep falling and avoiding the center of gravity because of the aforementioned, and so on.

By nature, human beings feel earthquakes as they occur even if they are minor and not devastating ones. Then how do they not feel the Earth's rotation around itself at a speed of a thousand miles an hour?

This is as you say that a bird over your chest feels your breath or some internal disturbances, and never feels when you sleep on your belly? Can you imagine such miraculous aspects of creation and such magnificent organization, yet we find many people disbelieve in Allah and the miracles of the prophets, such as the cleavage of the sea! Sadly enough, we, the Muslims, imitate them in ridiculing the flat Earth theorists, while at the same time, they do not mock those beliefs and views that actually deserve to be mocked and insult people's intelligence.

Some of the spherical Earth theorists have invoked that the planets look spherical, and so do the sun and the moon and use this as evidence that the Earth must be spherical as well. This, unfortunately, is negligence and lack of understanding, as what is the connective relationship between the planets, the sun, the moon, and the Earth? Who said the Earth is a planet, a star, or of the same genus of the celestial bodies? On what base have they concluded that? Let us imagine that there is a child who saw animals around him and noticed that they cannot speak, thus he concluded that man cannot speak as well, because animals could not do so. Or, he saw that animals walk on the four legs, so he concluded that all creatures walk on all four as well, although he sees them differ from each other. Can you imagine such conclusions? Really, the like of such conclusions force one to call them stupid.

If the Earth is moving, and so are the planets, why can we observe them although they are supposed to move at lightning speed? Let us imagine you are in your car when it is moving, but the ground is stable and the scenery around you is not self-moving—that is, your speed and the speed of what you look at are equal. However, if you are driving next to another car at different speeds, what will happen? You will find a great disparity, especially if you are moving in one direction and the observed object in the opposite direction. Why do not we find this disparity in the motions of the heavenly bodies, as their motions are fixed? If the bodies and the Earth are in constant motion and the Sun is also moving, and so is the galaxy, then why are their motions accustomed and fixed? If you say that the bodies move in the same movement of the Earth and in the same direction, then why do the locations of the stars differ in the poles?

I like to say that the concept of spherical Earth is old, not a modern one, as it has been prevalent for a long time. Therefore, it is wrong to claim the Earth's sphericalness is a contemporary theory but being an old theory does not necessitate it to be sound and valid. If validness is to be determined by the theories' oldness, the theory of the flat Earth is older than that of the spherical Earth. Satan is also older than most good people and had a view and opinion, but was he right about both of them?

In science, they cite Einstein's statement indicating that the simplest theories are the most accurate ones. I say to them, "Which is easier to understand and perceive: that the earth is spherical, floating in space, and revolving around the Sun, which is also moving, and so is the galaxy is moving, etcetera? Or, that the Earth is flat, the sky above us is like a ceiling, and the sun rises and sets?" The latter explains to you almost everything, even for a layperson who is not a scientist. Just imagine that you believe the Earth is flat for thousands of years and have all mass media support you and someone comes to you and says the Earth is spherical. It is said that the fruit is indicative of the type of the tree, and that the building indicates is an indicative of its base, so if you want to know some truth, look for the closer validations that are within reach, as they are usually the strongest and most persuasive. Do not strain yourself in discussing distant validations that are far away from you at distances calculated by mathematical equations and calculations the scientists themselves hardly understand.

The more reasonable and logical the evidence is and at all levels of understanding, the more powerful it becomes. There are no more powerful proofs that those which can be perceived by the senses. Praise is to Allah, all the senses indicate that the Earth is flat and stable, and the sun and the moon in constant motion, both diligently pursuing their courses. That is why people believe in that since ancient times. But when some scientists came up with complicated equations that no one understands but them, people were filled with respect and reverence for them, and therefore believed them. This is what usually happens to people, when they are confronted with something they are ignorant of, they feel fear of it. And when the feel fear, they would either fight it or surrender and acquiesce to it. Since scientists are human beings and with respectful personalities, fighting them was not the easiest choice for people, but rather they chose to acquiesce to them, and accordingly imitated them. That is how the people have become blind followers and imitators without being aware of the reasons for their imitation. It is only few people who follow the held views out of conviction, and after studying and understanding of all those theories produced by the scientific communities.

WHY IS ASTROLOGY MIXED UP WITH SCIENCE?

Unfortunately, some contemporary scientists of the scientific community resort to astrology and imagination more than to resort to the material objects and what can be tested physically and in front of their eyes. May I just for a moment pause here and say: Why do we mandate the flat earth theorists to prove to people the Earth's flatness while are the senses do prove that the Earth is flat? Is it not the followers of the spherical Earth theory who claim to have landed on the moon, and claim causes and reasons behind the so and so phenomenon or incident, and so forth? Thus, they are the party who shall prove their theory, not us. In a hadith attributed to the Prophet (peace and blessings of Allah be upon him), it is stated, "the onus of proof is upon the claimant."

Why do astronomers consider the moon to be orbiting the Earth while they say that the motion of the planets is only a perspective, which is the motion of the Earth and the Sun? Why they state these exceptions that have no convincing reason. They say that Jupiter revolves around itself every ten or five hours, so why don't people feel the change and cannot see the other side of the planet?

They say that seeing the disappearance of the ship over the horizon when being the seashore is a proof of the Earth's sphericalness. But to refute this weak and delusional proof, all you have to do is buy a camera with a powerful lens and look by yourself and you will discover that the ship has not disappeared and you can still see it with the camera, a matter which completely destroys the said proof. The reason why we see ships disappearing is the law of perspective and the limited extents of our visual acuity. Again, hold a book, like this one you are reading, move it away from you gradually now, and see what will happen? Its letters will get smaller, and if you move it away farther and farther, you will not be able to read the letters and if you get it away from you much farther, the letter will disappear. The reason behind that is the limited ability of the human eyes. Thus, if you use lenses, you will find yourself able to see and read the letters. Not only that, let us suppose you are on a ship and your friends are on other four ships, and each one of the ships heads in a direction and moves away from your ship, what will you notice? You'll see that the ships that move away from you northward, eastward, southward, westward are getting smaller, and that the ship you are on seems the biggest, right? Call a friend on each ship and ask them how they see the ships around them. They will all say that they see the other ships getting smaller and their own ship looks the biggest, why! Very briefly, because this is related to distance issues.

Also, if your trip is in a straight line and your ship in the middle and there are two ships in front of you at varying distances and two ships behind you at varying distances. So whenever the two ships in front of you move away farther, you will see them disappearing little by little. If you contact any person on them, he will say: your ship is the one that gets disappearing. This matter is understandable and simple. But because we used to believe their theories and get impressed with their experiments and arguments that the Earth is spherical, we have been no longer aware of what we say, and began to find difficulty in understanding what is simpler than it. This is Just as what happened with some polytheists when they believed in the multiplicity of gods and the existence of different gods, although this is more difficult to understand than to believe in only one God, the Creator of all creatures and the Lord of the Worlds. Allah has recorded their saying in the following Quranic verse: **"(they say) Has he made the gods [only] one God? Indeed, this is a curious thing"** [Sad:5]

I have searched for an experiment of building of an object that can spin a quarter of the Earth's speed, and float in space without being hung by a thread or a support. Considering technology has developed to the extent of establishing The International Space Station (ISS), they would have experimented something like this, right? Dear reader, if you have found any experiment like this, kindly send it to me, and I will be grateful to you.

It is known that before scientists come to results and conclusions, they first test the means. Then why do not they test if the Earth is spherical, revolving around itself, and around the sun, before they try to simulate the Earth exactly. Note that we can simulate the Earth and the heavens, and we do this daily in building houses. We can also test how water and liquids in general have a flat surface, but where are their experiments of stimulating the Earth without supporting it with threads, columns, bases, and the like?

What about magnets? If there is a magnet inside the core of the Earth, then how it does it work in the high temperature inside the Earth? Scientific data states that the temperature of the Earth's core is up to 6,000 degrees Celsius or 10,800 degrees Fahrenheit. How can the magnet work in such temperature? It should be known that this temperature is equal to the alleged temperature of the sun's surface. It is known that the magnets lose their magnetic strength if exposed to 80 degrees Celsius. If we assume the Earth was created four billion years ago and its temperature was more than the said figures, then four billion years are surely sufficient to cause any magnet to lose its strength. Therefore, the presence of Earth's gravity that pulls the moon and millions of tons of water is an astonishing matter and needs serious answers.

Finally, I would like to highlight a fact. Contemporary intellectuals used to say, "Whoever came after the Prophet is fallible." Besides, a number of those intellectuals publicize some errors of the Sharia scholars. When asked why they do so, they reply, "To remove the sacred aura surrounding them, so that the common Muslims will not follow them in such errors."

Suh good intentions are fine, but why do you not do the same thing with naturalists, or at least with philosophers who departed this world before the birth of Jesus, such as Aristotle, Plato, and others? Why is it hard for you to say, "The observations and expectations of so and so are wrong, and, therefore, everything that was based on such expectation and opinions is wrong as well." This is just as the case of a hadith scholar whose approach is weak, and therefore all his classifications of hadiths will be in doubt, right? Why don't you do the same thing with those who are held venerable by the Western community?

23

NIGHT AND DAY

I THINK NIGHT AND day are creatures other than the sun and the moon; they are "239awaher (essential)" and not "a'raad (contingent)," just as the sun and the moon. If you are wondering about the difference between 239awaher and a'raad, let me explain. The former refers to a thing that is already self-existent and has an essence, while the refers to a thing that does not have an essence and depends on something else in order to happen.

The simplest example is that a human being is of the 239awaher, but the human being's joy is of the a'raad, as it is contingent and depends on something else to happen. There is no self-existent thing called joy, and you cannot visualize the shape, essence, substance, or nature of joy. This is unlike the case when you say "human being," as you can recall people's known shape, appearance, and so on.

People say the night and day are a'raad and the sun is their *jawhar*, for the presence of the sun necessitate the presence of the a'raad, that is the day, and the absence of the sun necessitate the disappearance of the daylight and falling of the night. However, I say that night and day are of the *jawhar*, just as the sun and the moon are so. Besides, there is nothing that denies there is a correlation between them all, as the correlations Allah has made between many of his creatures. Allah, Exalted is He, said, **"And it is He who created the night and the day and the sun and the moon; all in an orbit are swimming"** [Al-anbiyāa:33]. Considering the apparent meaning of the verse, what immediately comes to the mind is that there are four things Allah has created. **"All in an orbit are swimming"** refers to the said four things: the night, the day, the sun, and the moon. I do not find strong evidence for those who said it only refers to the sun and moon. The verse also means the night has an orbit where it spins, and the same applies to the day, the sun, and the moon. The fact that both night and day have orbits corresponds to the saying of Allah, the Almighty, **"He created the heavens and earth in truth. He wraps the night over the day and wraps the day over the night and has subjected the sun and the moon, each running [its course] for a specified term. Unquestionably, He is the Exalted in Might, the Perpetual Forgiver"** [Az-Zumar:5].

Whoever ponders on the Quranic verses finds that when Allah, Exalted is He, uses the verb "created," 240*awaher* are stated after it, such as the heavens, earth, humankind, sun, moon, and other *240awaher*, and will not find *a'raad*. So, what makes them say that night and day are of the *a'raad*?

If I am asked about Allah's Saying, **[He] who created death and life to test you [as to] which of you is best in deed – and He is the Exalted in Might, the Forgiving"** [Al-Mulk:2], I will reply that the answer is in Allah's Saying, **"Wherever you may be, death will overtake you, even if you should be within towers of lofty construction"** [An-Nisaa:78].

What shows me that life and death are not of the *a'raad,* and it is stated in hadith reported by Abu Sa'id al-Khudri, may Allah be pleased with him, that the Prophet (peace and blessings of Allah be upon him) said, "Death would be brought on the Day of Resurrection in the form of a white-coloured ram. It would be said to the inmates of Paradise: Do you recognize this? They would raise up their necks and look towards it and say: Yes, it is death. Then it would be said to the inmates of Hell-Fire: Do you recognize this? And they would raise up their necks and look and say: Yes, it is death. Then command would be given for slaughtering that and then it would be said: O inmates of Paradise, there is an everlasting life for you and no death. And then (addressing) to the inmates of the Hell-Fire, it would be said: O inmates of Hell-Fire, there is an everlasting living for you and no death. Allah's Messenger then recited this verse pointing with his hand to this (material) world: And warn them, [O Muhammad], of the

Day of Regret, when the matter will be concluded; and [yet], they are in [a state of] heedlessness" [Maryam:39].

Those who held false views are also in a state of heedlessness in this worldly life and "do not believe." Allah know best.

There is also a similar Quranic verse to the aforementioned one where Allah, Exalted is He, said, **"It is not allowable for the sun to reach the moon, nor does the night overtake the day, but each, in an orbit, is swimming"** [Yasin:40]. Allah also said, **"[He is] the cleaver of daybreak and has made the night for rest and the sun and moon for calculation. That is the determination of the Exalted in Might, the Knowing"** [Al-an'ām:96], and, **"And He subjected for you the sun and the moon, continuous [in orbit], and subjected for you the night and the day"** [Ibrahim:33]. This is besides Allah's Saying: **"And He has subjected for you the night and day and the sun and moon, and the stars are subjected by His command. Indeed, in that are signs for a people who reason"** [An-Nahl:12]. Allah mentioned the sun and the moon, and in the same verse, He mentioned the day and the night as well, indicating that they are four distinct, separate four things. Allah also said: **"And of His signs are the night and day and the sun and moon. Do not prostrate to the sun or to the moon, but prostate to Allah, who created them, if it should be Him that you worship"** [Fuṣṣilat:37], this verse indicates that those four are of Allah's signs as He said: **"the night and day and the sun and moon."**

Allah also stated in the chapter of As-shams what immediately indicates, from the verses' apparent meaning, that by the day, it displays the sun's brightness: **"(1) By the sun and its brightness, (2) And [by] the moon when it follows it, (3) And [by] the day when it displays it, (4) And [by] the night when it covers it"** [Ash-Shams:1–4]. The verses included matters that are related to the sun, such as stating the moon follows it, the day shows it, and the night enshrouds it. This also means the functions of the day is to display the sun and that of the functions of the night is to conceal of the sun. Whoever ponders on only the cases of night mentioned in the thirtieth part, "Juzu Amma" will find six cases by which Allah has sworn. Besides, it is very difficult to consider the night as being merely contingent based on these verses whose apparent meaning indicates that the night is a self-existent entity. Allah said, **"By the Night as it conceals (the light)"** [Al-Layl:1], **"And [by] the night when it passes"** [Al-Fajr:4], **"And [by] the night when it covers with darkness"** [Ad-Duha:2], **"And [by] the Night as it dissipates"** [At-Takwir:17], **"And [by] the night and what it envelops"** [Al-Inshiqāq:17], and **"And [by] the night when it covers it"** [Ash-Shams:4]. These are the meant cases whose number is six, just as the days of the creation of the heavens and the Earth, which prove that the night is a self-standing and moving sign. Besides, it floats as Allah has stated and is not the result of the rotation of the Earth around the sun, as claimed.

In addition to the abovementioned, there are multiple verses where the night and day are mentioned before the sun and moon. Let us ponder on the following verses.

"Indeed, your Lord is Allah, who created the heavens and earth in six days and then established Himself above the Throne. He covers the night with the day, [another night] chasing it rapidly; and [He created] the sun, the moon, and the stars, subjected by His command. Unquestionably, His is the creation and the command; blessed is Allah, Lord of the worlds" [Al-aʿrāf:54.

"Do you not see that Allah causes the night to enter the day and causes the day to enter the night and has subjected the sun and the moon, each running [its course] for a specified term, and that Allah, with whatever you do, is Acquainted?" [Luqman:29].

"He causes the night to enter the day, and He causes the day to enter the night and has subjected the sun and the moon – each running [its course] for a specified term. That is Allah, your Lord; to Him belongs sovereignty. And those whom you invoke other than Him do not possess [as much as] the membrane of a date seed" [Fatir:13],

"He created the heavens and earth in truth. He wraps the night over the day and wraps the day over the night and has subjected the sun and the moon, each running [its course] for a specified term. Unquestionably, He is the Exalted in Might, the Perpetual Forgiver" [Az-Zumar:5]

"And of His signs are the night and day and the sun and moon. Do not prostrate to the sun or to the moon, but prostate to Allah, who created them, if it should be Him that you worship" [Fuṣṣilat:37].
This succession cannot occur haphazardly, and there must be great wisdom behind it. It appears to me that the creation of the night and day preceded the creation of the sun and moon. Praise is to Allah, what made me reassured about this viewpoint is that such a fact has been stated clearly in the Torah—namely in Genesis, which is the first book thereof—where it is stated that God created the day and night on the first day, while he created the sun and moon on the fourth day. The believers in the modern science have used this fact to claim the Torah to be a scientifically outdated book. Yet, it seems to me that the Quran also confirms this fact.

Glorified is Allah, whenever I search in the Quran or the Torah regarding the creation of the heavens, the Earth, the night, the day, the sun, and the moon, I find marvelous concordance between them that makes me echo the sentiments of Negus, ruler of Ethiopia, "This (the Quran) and what Prophet Moses brought came from the same source." As for the alleged misinterpretation and distortion of the Torah, even if we accept that its letters and words have been distorted, the parts that are related to the creation of the heavens and the Earth have not been distorted.

Let us ponder on Allah's Saying, "(1) **Are you a more difficult creation or is the heaven? Allah constructed it. (2) He raised its ceiling and proportioned it. (3) And He darkened its night and extracted its brightness. (4) And after that He spread the earth"** [An-nāziʿāt:1–4]. Contemplate the connection between the night, day, and heaven, and then contemplate again how Allah mentioned them before extending the Earth. This indicates that the night and day are self-existent entities and not mere signs of sunrise or sunset. Furthermore, it seems from these verses that the night and day are greater, more wondrous, and more indicative in terms of manifesting Allah's ability than the sun and the moon. Perhaps this is the reason we find that night and day have been mentioned in numerous verses, while the sun and moon have been mentioned very few times; maybe in just one or two verses, if my memory does not deceive me.

It should also be noted that Allah's sign of day and night may include the angels. Allah, Almighty, said, **"They exalt [Him] night and day [and] do not slacken"** [Al-anbiyāa:20]. Ibn Jarir al-Tabari said commenting on the meaning of this verse, "They are the angels who glorify Allah's Praise and nor do they ever flag or intermit to do so." It is well known to people that angels do not live on a spherical Earth, but they are mostly in heaven and, that is why, Allah said, **"Surely, those who are with your Lord (angels),"** and this is what comes immediately to one's mind. Also, there is the well-known hadith where the Prophet said, "I see what you do not see, and I hear what you do not hear. The heaven is creaking and it should creak, for there is no space in it the width of four fingers but there is an angel there, prostrating to Allah."

Where is the night and day in the space, according to the modern science? And which sun is responsible for the night and day in the heaven? The aforementioned verses indicate that night and day are two great signs that are separate from the signs of the sun and moon, and there is nothing denies the fact that there is a connection between all of them. Thus, night and day are of Allah's miraculous signs and the indication of their outset is sunset and sunrise. In my opinion, the main purpose of the sun and the moon is to know the number of years and calculation, and they serve to signal the beginning of the day and night. Perhaps, therefore, we find Allah's saying in the Quran, **"And We have made the night and day two signs, and We erased the sign of the night and made the sign of the day visible that you may seek bounty from your Lord and may know the number of years and the account [of time]. And everything We have set out in detail"** [Al-Israa:12]; and, **"It is He who made the sun a shining light and the moon a derived light and determined for it phases – that you may know the number of years and account [of time]. Allah has not created this except in truth. He details the signs for a people who know"** [Yūnus:5]; and, **"The sun and the moon follow courses (exactly) computed"** [Ar-Rahman:5]. This is apart from what is stated in the first chapter of the book of Genesis, the 14th verse, *"And God said, Let there be **lights** in the firmament of the heaven to divide the day from the night; and let them be for signs, and for seasons, and for days, and years."*

While most Quranic verses regarding day and night clarify that the night is for rest and the day is for making a living and seeking Allah's bounty, their functions are known, as are the functions of the sun and moon.

As I said earlier, there is nothing denies that the limited lighting is one of the sun's functions, and that the limited illumination is one of the moon's functions. Furthermore, these verses contain numerous pieces of scientific information, and I have a lot to say about that, but this book cannot accommodate all of it. Besides, I fear the readers may not comprehend much of what I say, and this is not out of personal arrogance, God forbid, but out of fear of drifting away from the purpose of the book, as a detailed explanation will take too long.

Allah, Exalted is He, said, **"Indeed, your Lord knows, [O Muhammad], that you stand [in prayer] almost two thirds of the night or half of it or a third of it, and [so do] a group of those with you. And Allah determines [the extent of] the night and the day. He has known that you [Muslims] will not be able to do it and has turned to you in forgiveness, so recite what is easy [for you] of the Quran. He has known that there will be among you those who are ill and others traveling throughout the land seeking [something] of the bounty of Allah and others fighting for the cause of Allah. So recite what is easy from it and establish prayer and give zakah and loan Allah a goodly loan. And whatever good you put forward for yourselves – you will find it with Allah. It is better and greater in reward. And seek forgiveness of Allah. Indeed, Allah is Forgiving and Merciful"** [Al-muzamil:2]. Ponder on Allah's Saying, **"Allah determines [the extent of] the night and the day,"** and that Allah did not mention the sun and the moon. Thus, if the night and day are of the *a'raad* and not of the *awaher*, then the reason for mentioning day and night in this verse would be incomprehensible. Moreover, there are several verses proving that the night is a self-standing and non-contingent entity.

Allah said, **"(71) Say, 'Have you considered: if Allah should make for you the night continuous until the Day of Resurrection, what deity other than Allah could bring you light? Then will you not hear?' (72) Say, 'Have you considered: if Allah should make for** you the day continuous until the Day of Resurrection, what deity other than Allah could bring you a night in which you may rest? Then will you not see?' (73) And out of His mercy He made for you the night and the day that you may rest therein and [by day] seek from His bounty and [that] perhaps you will be grateful" [Al-qaṣaṣ:71–73]. Why didn't Allah say in these verses that he Holds the sun or anything that is related to the sun? Do these verses not indicate that night and day are different from sun and moon, and that they are not dependent on sunrise or the sunset?

If we say that the day is something and the sun is something else, a question will pop into the mind, "Does the day exist without the sun?" Surely, Allah knows best, but the answer I consider more probable is, "Yes, and the sun is the sign of the day, and the start of the daytime, and the day displays the sun." We have lived many years, thanks to Allah, and experienced solar eclipses, and I have never heard of a person who saw the day darken when the phenomenon occurred. Then, why do we always see the day with the sun? Well, let me accept this view and say this is a fact, although do I see the daylight in the sky prior to the sunrise on a daily basis. Of course, the daylight does gets stronger after sunrise.

Anyway, if we say that there is a relationship between the day and the sun, as it is said that the day and the sun are always together, then how do we differentiate between them, since we say that the day is of the 246awaher and so is the sun? This is possible. Let us imagine the day in the form of a full circle in which the sun rotates downward, upward, or within its orbit, so wherever the sun goes, the day is there with it, because the sun is within the circle of the day. Do you know the Chinese yin-yang symbol? If you look at it, you will find a black point in the white-colored part and a white point in the black-colored part. Let us imagine that the day is like the white-colored part and the sun is like the black point within it, and both have equal movement and speed. This is how they both are of the 246awaher but of different natures, and due to the accurate and systematic movement, we always find them accompanying each other. Try to draw this, then judge yourself if this is possible. I say it is possible. Moreover, it tells me that this concept is closer the mark.

There is another point. I think darkness or the night was created before day, or that it is a greater sign than the day. Note that greater here does not necessarily mean it is better than the day. It is just that the night is mentioned in the Quran about ninety-two times while the day is mentioned about fifty-seven times. Also, we often find in the Quran that when night and day are mentioned together, the night is mentioned first. For example, Allah said, **"Indeed, in the creation of the heavens and the earth and the alternation of the night and the day are signs for those of understanding"** [āl ʿimʾrān:190]; and, **"And it is He who created the night and the day and the sun and the moon; all [heavenly bodies] in an orbit are swimming"** [Al-anbiyāa:33]; and, **"Indeed, your Lord is Allah, who created the heavens and earth in six days and then established Himself above the Throne. He covers the night with the day, [another night] chasing it rapidly; and [He created] the sun, the moon, and the stars, subjected by His command. Unquestionably, His is the creation and the command; blessed is Allah, Lord of the worlds"** [Al-aʿrāf:33]. Also, Allah said, **"It is He who made for you the night to rest therein and the day, giving sight. Indeed in that are signs for a people who listen"** [Yūnus:67]; and, **"And We have made the night and day two signs, and We erased the sign of the night and made the sign of the day visible that you may seek bounty from your Lord and may know the number of years and the account [of time]. And everything We have set out in detail"** [Al-Israa:12]. This is apart from many other verses where Allah mentions the night before the day.

Yes, the night has many virtues, but this also indicates that the night originated before the day, as the following verse refers, **All praises and thanks be to Allah, Who (Alone) created the heavens and the earth, and originated the darkness and the light, yet those who disbelieve hold others as equal with their Lord** [Al-anʿām:1]. The Torah, which mentions that darkness was at the beginning till God made the light or the day, mirrors this belief. It is stated in the Torah, in the book of Genesis, in the first verses of the first chapter, *"(1) In the beginning God created the heaven and the earth. (2) And the earth was without form, and void; and darkness was upon the face of the deep. And the*

Spirit of God moved upon the face of the waters. (3) And God said, Let there be light: and there was light. (4) And God saw the light, that it was good: and God divided the light from the darkness. (5) And God called the light Day, and the darkness he called Night. And the evening and the morning were the first day."

It is as if darkness is the mother of the day, or, shall we say, the mother of the creatures that came after the creation of the heavens, the Earth, and this human existence. It is stated in the prophetic hadith, "Indeed Allah, the Blessed and Exalted, created His creation in darkness, then He cast His Light upon them . . . " This shows me that the night or darkness had originated before the day, the sun, and the moon. Allah, Almighty said, **"And a sign for them is the night. We remove from it [the light of] day, so they are [left] in darkness"** [Yasin:37], as if the day is like a cover of leather on the night and it is stripped of it. Also, we find Allah said about the night, **"By the night when it covers"** [Al-layl:1]; and, " . . . He brings the night as a cover over the day, seeking it rapidly . . . " [Al-a'raf:54].

Ponder on Allah's saying, **"By the night when it covers"** [Al-layl:1]. Cover here is the translation of the Arabic word mentioned in the verse, "yagh-sha," When "*ghasha* his head" is said in Arabic, it means that he "covered his head," and "*ghasha*" means that thing is "covered." Also, when it is said that a man "*ghasha*" his wife, it means he was above her, having sexual intercourse with her, as if he is her cover. Allah said, **"It is He Who has created you from a single person (Adam), and (then) He has created from him his wife [Hawwa (Eve)], in order that he might enjoy the pleasure of living with her. When he 'taghash-shaha' (i.e. had sexual relation with her), she became pregnant and she carried it about lightly. Then when it became heavy, they both invoked Allah, their Lord (saying): 'If You give us a Salih (good in every aspect) child, we shall indeed be among the grateful.'"** [Al-a'raf:189].

Also consider the Quranic verse, **"When that 'yaghsha" (i.e. covered)' the lote-tree which 'yaghsha' (i.e. did cover) it!"** [An-Najm:16]. Here too, "*yaghsha*" means to cover and shroud. Allah also said, **"Or [they are] like darknesses within an unfathomable sea which 'yaghshaho' (i.e. is covered) by waves, upon which are waves, over which are clouds**

– darknesses, some of them upon others. When one puts out his hand [therein], he can hardly see it. And he to whom Allah has not granted light – for him there is no light" [An-nur:40].

From the abovementioned verses, we infer that the covering is usually from above, like a cover or veil. This applies to the night, as it covers and shrouds the day. Perhaps this is why the higher up a person goes toward the sky, the darker the world seems, because he is about to come out of the scope of the day and enter the scope of the great night, and Allah knows best.

24

THE SUN

I NSTEAD OF REPEATING what has been said before concerning the matter of the rotation of the sun, and in order to save time and effort, it is better for me to state the opinion of Sheikh Muhammad bin Salih al-'Uthaymin (may Allah have mercy on him). He provided a clear-cut answer from which I am going to quote what is related to the sun.

"As for our opinion on the rotation of the sun around the Earth by which the succession of night and day occurs, we adhere to the apparent meanings of the texts of the Quran and the Sunnah that indicate that the sun rotates around the Earth, a matter by which the succession of night and day occurs. This is in order to have conclusive evidence to be a testament for us based on the apparent meanings of the Quran and Sunnah. This is what should be done as it is obligatory for the believer to stick to the apparent meanings of the Quran and Sunnah concerning such issues and others. Of the proof that the sun rotates around the Earth, which results in the succession of the night and day is Allah's saying: '**And [had you been present], you would see the sun when it rose, inclining away from their cave on the right, and when it set, passing away from them on the left**' [Al-Kahf:17]. Here are four verbs whose actions return to 'the sun,' they are: 'rose,' 'inclining away,' 'set,' and 'pass away.' If the succession of night and day is caused by the rotation of the earth, then it should be said in the verse 'you would see the sun when the earth surface is shown to it' or the like. Also, it is reported that the Prophet said to

Abu Dharr, 'O Abu Dharr! Do you know where this (sun) goes?' Abu Dharr said, 'Allah and His Messenger know better.' The Prophet said, 'Indeed it goes to seek permission to prostrate, so it is permitted.' And it is as if it has been said to it, 'Rise up and emerge out from the place of your setting, and it will rise from the place of its setting.'

"The verbs 'goes, rising up, emerging out, and setting,' which return to the sun, show that the succession of the night and day is the result of the rotation of the sun around the Earth. As for what is stated by the contemporary astronomers in this regard, it has not reached the degree of complete certainty for us and thus we should not, for its sake, ignore the apparent meanings of the Quran and Sunnah. Moreover, we advise whoever is assigned to teach geography to clarify to the students that the apparent meanings of both the Quran and the Sunnah that the succession of night and day is due to the rotation of the sun around the Earth, not the other way around.

"If a student says, 'which one shall we adhere to: the apparent meanings of the Quran and the Sunnah or what they claim that it is a certainty?' The answer will be, 'We shall adhere to the apparent meanings of the Quran and Sunnah because the Quran is the wording of Allah, Exalted is He, Who is the Creator of the whole universe, the whole world with all of its contents and conditions, movement and stillness.' Besides, surely the words of Allah, Almighty, are the most truthful and clear-cut, as He sent down the Quran as clarification of all things, and He has stated that He has clarified everything for His slaves to not turn away from the right path. As for the Sunnah, it is the wording of the Messenger sent by the Lord of the worlds, and he is the most knowledgeable person of Allah's rulings and actions. He also did not utter a word out of his own inclinations but it is out of a divine revelation, as such issues cannot be received from any other source except a divine revelation. I think, and Allah knows best, that there will be a time when the views of contemporary astronomers will be destroyed, just as what happened with Darwin's view on the human origin and evolution."

Sheikh Ibn 'Uthaymin stated the following about the sun's rotation around the Earth:

"The apparent meanings of the Shariah proofs prove that it is the sun which rotates around the Earth, and such rotation results in the succession of the night and day on the Earth surface. Thus, we shall not disregard the apparent meanings of these proof except if there is a more powerful proof that justifies for us to interpret those proofs with meanings other than the apparent ones. Of the proofs that the sun rotates around the earth and such rotation causes the succession of the night and day:

1. Allah said, stating what Ibrahim was saying to the one who argued with him about Allah, **'Abraham said, 'Indeed, Allah brings up the sun from the east, so bring it up from the west''** [Al-baqarah:258], so the fact that the sun is brought up from the east is an apparent proof that it is the one which rotates around the Earth.

2. Allah also said about Ibrahim, **'And when he saw the sun rising, he said, 'This is my lord; this is greater.' But when it set, he said, 'O my people, indeed I am free from what you associate with Allah''** [Al-an'ām:78]. It shall be noted that it is mentioned the sun is the one which set not that Ibrahim went away until it disappeared and thus if the Earth is one which rotates, then it should be said instead 'when he went away from it until it disappeared.'

3. Allah said, **'And [had you been present], you would see the sun when it rose, inclining away from their cave on the right, and when it set, passing away from them on the left'** [Al-Kahf:17]. As is shown from the verse, the verbs of inclining away and of passing away return to the sun, and this is proves that it is the sun which rotates around the Earth, because if is the earth then it should be said instead 'their cave inclined away from them.' Besides, making rising and setting relate to the sun evidences that the sun is the one which rotates, even if this is a less powerful proof than Allah's words **'inclining away from'** and **'passing away.'**

4. Allah said, **'And it is He who created the night and the day and the sun and the moon; all [heavenly bodies] in an orbit are swimming'** [Al-anbiyāa:33]. Ibn 'Abbas, may Allah be pleased with both of them, said, 'they spin in a whorl like the spindle whorl – this is a popular statement attributed to Ibn Abbas.'

5. Allah said, 'He covers the night with the day, [another night] chasing it rapidly' [Al-a'rāf:54]. In this verse, the night is depicted as chasing or seeking the day in a rapid and urgent way. It is known that the night and day follow the sun.

6. Allah said, 'He created the heavens and earth in truth. He wraps the night over the day and wraps the day over the night and has subjected the sun and the moon, each running [its course] for a specified term. Unquestionably, He is the Exalted in Might, the Perpetual Forgiver' [Az-Zumar:5]. Allah's saying 'He wraps the night over the day and wraps the day over the night' is a proof of the spinning of the day and night around the earth, and if the earth is the one which spins around them, then it shall be said instead 'He wraps the earth over the night and day.' Allah's saying 'each running [its course] for a specified term,' which clarifies what is stated before it, is a proof that the said running of the sun and moon is real and spatial because subjecting a moving object to do motion is more reasonable than subjecting a stable object that does not move.

7. Allah said, '(1) By the sun and its brightness, (2) And [by] the moon when it follows it' [Ash-Shams:1–2]. The word 'follows' is evidence of their spinning and rotating around the Earth, and if it is the Earth that spins around them, then moon will not always follow the sun, but will follow it sometimes, and sometimes the sun will follow it as the sun is greater than the moon. Using this verse as evidence needs to be deeply considered.

8. Allah said, '(38) And the sun runs [on course] toward its stopping point. That is the determination of the Exalted in Might, the Knowing. (39) And the moon – We have determined for it phases, until it returns [appearing] like the old date stalk. (40) It is not allowable for the sun to reach the moon, nor does the night overtake the day, but each, in an orbit, is swimming' [Yasin:38–40]. Mentioning that the sun runs due to the determination of Allah, the All-Mighty, the All-Knowing, indicates that it is real running that is determined wisely and that is why it results in the

succession of the night and day and of the seasons. Determining phases for the moon also indicates that it moves among them and if it's the Earth that spins, then such phases should be determined for it from the moon and not for the moon. Negating that the sun can reach the moon and that the night can overtake the day indicates the hasty motion of the sun, moon, night and day.

9.   The Prophet (peace and blessings of Allah be upon him) said to Abu Dharr (may Allah be pleased with him), 'O Abu Dharr! Do you know where this (sun) goes?' Abu Dharr said: 'Allah and His Messenger know better.' The Prophet said, 'Indeed it goes to seek permission to prostrate, so it is permitted.' And it is as if it has been said to it, 'Rise up and emerge out from the place of your setting, and it will rise from the place of its setting.' The Prophet's saying 'Rise up and emerge out from the place of your setting' clearly indicates that the sun rotates around the Earth and such rotations causes the sunrise and sunset.

10. The various hadiths where the sunrise, sunset, and sun's passing of its zenith are mentioned clearly indicate that it is the sun rotates around the Earth not the other way around.

Perhaps there are other proofs that I do not remember right now, but what I have mentioned is sufficient in this regard. And Allah is the One Who grants success!"

As you we have seen, the proofs he stated are so powerful that I do not see the need to add anything to his words. Yet, I would like to add some verses where the motion of the sun and the moon is mentioned and that are suitable to be proofs in this regard, namely for the motion of the sun. Of these verses, **"It is Allah who erected the heavens without pillars that you [can] see; then He established Himself above the Throne and made subject the sun and the moon, each running [its course] for a specified term. He arranges [each] matter; He details the signs that you may, of the meeting with your Lord, be certain"** [Al-raʿd:2]. Also, Allah saying, **"Do you not see that Allah causes the night to enter the day and causes the day to enter the night and has subjected the sun and the moon,** each running [its course] for a specified term, and that Allah, with whatever you do, is Acquainted?" [Luqman:29]; and, "He causes the night to enter the day, and He causes the day to enter the night and has subjected the sun and the moon – each running [its course] for a specified term. That is Allah, your Lord; to Him belongs sovereignty. And those whom you invoke other than Him do not possess [as much as] the membrane of a date seed" [Fatir:13].

These verses, in addition to the verse mentioned by Sheikh Ibn 'Uthaymin, where Allah informs us that the sun runs, besides reminding us of the grace He has endowed us withby saying, **"And He subjected for you the sun and the moon, continuous [in orbit], and subjected for you the night and the day"** [Ibrahim:33], all indicate the motion of the sun. I am going to explain this in more detail in the following pages.

Do contemplate Allah's saying, **"He subjected for you,"** which indicates that Allah created the sun and made spins so that we benefit from such matter. So, if the Earth is one of the sun's results and if it rotates around the sun, and as we are a mere atom in this cosmos, then how people will understand Allah's saying, **"He subjected for you."** Yes, I can find some justifications and some interpretations that change the apparent meaning that conform with modern science. Yet, will whatever I can find be in conformity with the intended meaning of the verse?

Also, in the Quranic verse, **"Establish prayer at the decline of the sun [from its meridian] until the darkness of the night and [also] the Qur'an of dawn. Indeed, the recitation of dawn is ever witnessed"** [Al-Israa:78], the exegetes of the Quran, such as of Ibn Mas'ūd, Mujahid, and Ibn Zaid, stated that Allah's saying **"at the decline of the sun [from its meridian]"** refers to sunset. This is also stated by Ibn 'Abbas, Ibn 'Umar. This interpretation was favored by Al-Hasan, Ad-dahaq, Qatada, and Ibn Jarir At-Tabari. This is a proof of the motion of the sun. Also, in Allah's saying: **" . . . and exalt [Allah] with praise of your Lord before the rising of the sun and before its setting"** [Taha:130], and **"So be patient, [O Muhammad], over what they say and exalt [Allah] with praise of your Lord before the rising of the sun and before its setting"** [Qaf:39], the verbs "rising" and "setting" return to what? To the sun, and this is due to its motion.

Regarding the movement of the sun, there are more than ten proofs in the Sunnah related in different ways, needless to be mentioned one after another, as the mentioned verses suffice. Additionally, the movement of the sun and evidence of such movement are in the two parts of the Bible—the Old and the New Testament—exceed fifty.

Here, a question may be raised. "Why has not Allah mentioned anything about the sun's activities or its effect on Earth as is proved today of its rotation around the earth due to its pull gravity." Allah mentioned the wisdom behind the creation of stars and meteors that monitor "Shatateen" (devils and demons), as well as the moon and the sun, which have been reported to be created in order to know the number of years and calculation.

Why have we not found even a part of verse that says the Sun attracts the Earth or does such-and-such for the wellbeing of humanity? Surprisingly, some believe their senses when it comes to the tangibles, yet they do not trust these very same senses, let alone that they see that the sun is moving in constellations and phases, leaning toward sometimes the North and toward the South at other times, just like the Moon, besides that the sun is nearly the same size of the moon, the Earth is stationary, and others.

Why do they not trust their senses when it comes to the issue of spherical Earth? Why do they insist on disregarding their senses for the sake of the reports of scientific periodicals on the spherical Earth? Assumingly they do not trust the senses in this field, I wonder how they can trust the output they have chosen for themselves in the same field? Have you scrutinized their view? You would conclude that it could be a counterclaim against them. It is unbecoming to turn their backs to their senses or the sacred Revelations for mere, baseless scientific findings.

WHERE DOES THE SUN GO IN THE FLAT EARTH?

The answer is that since the sun is smaller than the Earth and is a great distance away from us, it looks like it disappears underground. Then, it appears closer when it shines, and thus looks as if it rises up as a result of the vision from a certain angle and the law of perspective. Otherwise, it has a circular motion over the flat ground ceaselessly.

In this context, Allah, the Almighty, says, **"And He subjected for you the sun and the moon, continuous [in orbit]"** [Ibrahim:33].

In his interpretation of the preceding verse, Qurtubi said, "to the repair of the plant and others. The Arabic word 'Da'bein' linguistically indicates performing an ongoing action, i.e. they are in continuous motion in compliance to Allah's Commands. In other words, they will move continuously till the Day of Resurrection." The same view was held by Ibn Abbas, and Ibn Kathir said, "they move consistently and continuously; day and night."

In the same vein, Al-Zamakhshari said, "They are diligent in motion and illumination to dispel darkness and repair any corruption in earth, bodies and plants." Fakher Al-Razi holds that 'Da'bein' linguistically means to have the action done on a steady base, that is, to be consistently persistent as mentioned in Allah's wording, **[Joseph] said, "You will plant for seven years consecutively"** [Youssef:47].

Some exegetists hold that **"Da'bein"** means that they persist in their motion, illumination and influence in dispelling darkness, and in the repair of plants and animals. Anyway, this verse proves that the sun and the moon have a subjective movement, not the movement of the Earth, and that is what Al-Alwsi alluded to in his book, *Tafsir Rouh Al-Manai* (*Interpretation of the Spirit of Meanings*), saying, "The apparent meaning of the verse is a proof of their own movement. There is a beautiful narration relating to the verse meaning in which Hassan Al-Basri says, 'If the sun sets, it spins in the orbit of the sky which is anticlockwise of the qiblah direction, until it returns to the east from which it rises up and moves in the sky from east to west. Then it returns to horizon i.e. it moves anticlockwise of the Qibla direction to the east. Accordingly, it rotates in its created orbit and so is the Moon.'"

Since the sun is much smaller than the Earth, it spins in circular motion, its light has boundaries, and it does not cover the whole earth, so a part of the Earth is dark and the other part exposed to the sun is lit, and so on. The query of the change of the size of the sun when passing from east to west, this can be ascribed to the change of vision angle.

Imagine standing in a street and watching a car coming from the horizon. The car will seem very tiny, and then it seems gradually get bigger until it reaches to you. At that moment, you see it bigger than it was at the horizon.

Tell me, has the size of the car changed, or is it the same? Certainly, the size has not changed, but the reason the size varies according to the distance of vision can be explained in the realm of what has become known today as perspective. Thanks to this phenomenon, you see that the size varies, which is the same with the sun. Analogical is the case with the sunrise and the sunset, as the brain is affected by the perspective, and you see the sun grow on the horizon. This interpretation is used for the Earth's spherical theory as well as for the flat Earth, hence it is not problematic.

Nonetheless, the view that the Earth is flat adopts that "sky atmosphere" on the horizon acts as a magnifier because of humidity. Though the sky is clear and there are no obstacles, it seems smaller. This is known to all. These obstacles and water molecules in the air, along with other elements, cause this effect. It is true that one views the sun at sunset with a bigger size, but its light is less intense and less effective. This is what happens when light is refracted through a dense object, especially if this dense intermediate body contains water.

Let us take another example. If you were standing on the street and looking at cars coming in sight, the size of the lights of the car closest to you will surely be larger than the cars behind it, but if you focus on one of the farther cars as it approaches you, you will find that its lights are the same as the others. Another example is to use a flashlight and direct its beams to a wall in your room. Stand close to the wall. How do see the flashlight reflection? Almost the size of the flashlight opening, right? Well, what happens if you turn away and still point the light at the wall and back off a few steps? The reflection of light grows.

Similarly, try to stick your face onto the mirror and mark the circumference of your face on the reflection of your image on the mirror. Then move away and observe. You will see that the reflection of your image grows smaller than the marks you put on the mirror, right? These simple experiments help us understand how perspective plays its part. You should note that the quality of the beams source plays a pivotal role, as well as their strength and distance. We can observe these things as they are smaller than the Sun and closer to us, and there are hindering factors between us such as humidity in the atmosphere that affects our perception of the sun when it sets on the horizon.

Some may try to outsmart us by saying that the sun cannot be seen at night in the spherical Earth theory because the Earth is spherical and it revolves around itself, but what is the interpretation of those who embrace the theory of flat Earth? Why cannot we see the sun at night?

The reason is simply because the sun is spinning in its orbit three thousand miles or more above the Earth, but certainly not the number close to the scientific community, which is more than two million miles away, and your sight is limited to no more one hundred miles. If you look at the sky to the best of your visual abilities, you won't be able to look at the sun after it is gone. Apart from the limited vision, there are many factors that prevent you from seeing.

In addition, the sun is much smaller than the Earth, and it spins above Earth in its orbit, together with the fact that the day is lit in a spot of flat land, being dark in another spot. Imagine four hundred and one men, with two hundred of them one side, two hundred on the other, all forming two long, horizontal lines, and the man in the middle holds a small umbrella over his head. Will the small umbrella shade the men standing on the far right? Rather, can the one standing far right look at the umbrella? Will the umbrella shade the men standing furthest left? Can those who stand furthest left see the umbrella? This is a simple example. Though the difference is big, it explains why people can't look at the sun at all times. Therefore, there is a spot on the Earth that has sunlight where the sun exists and a spot that is dark, as the sun does not reach it, exactly like the case with the umbrella that failed to provide shade to four hundred men.

As the Sun spins circularly over the flat Earth, it moves away and approaches us vertically and horizontally, causing the variation of the seasons and the length of the day or night. Here, I would like to convey what is narrated from Ibn 'Umar, may Allah be pleased with them:

"If the sun spins in the same way, none on earth would have been benefited from it a bit. Yet, it spins in a circular motion in summer and spins otherwise in winter. If it spins in summer the way it does in winter, heat would have perished all; if it spins in winter the way it does in summer, cold would have perished all.

"In order to prove that the sun rotates in a circular motion over the globe in the middle of the earth, suppose you are in the middle and have a painting. Draw the path of the sun, as you draw the path of the Sun, you will be sure that you have drawn a half circle, had you asked a friend on the opposite side to draw half a circle opposite your circle.

"If you conjoined the two paintings together, you would notice a circular motion i.e. the path of the Sun is circular above the flat Earth, which makes no sense if the earth is spherical orbiting around its axis and orbiting the Sun. You can also test the proximity of the sun in March, the 21st day, when the day and night are equal by being present at the equator watching your own shadow. It is supposed that the shadow to be at the west during sunrise and until noon, and at the right during sunset, but why does not this happen?

"Why does the shadow move in a circular motion as in previous experience?

"This indicates two things, that the sun is much smaller than they claim and is closer to us as well.

"Perhaps I am a little moved away from the question of the disappearance of the sun, but in a nutshell and lastly, it suffices to say that when the night falls, it obscures the sun. Allah, The Almighty, says in His book:

'**(1) By the sun and its brightness (2) And [by] the moon when it follows it (3) And [by] the day when it displays it (4) And [by] the night when it covers it.** [Ash-Shams:1–4]'

"The day displays the sun and the night covers it and it is true that the sun would be in another spot in its orbit, yet with the nightfall, one can't, by no means, look at the sun at night unless one goes to the bright spot of the sun as the night has not covered it yet. Nonetheless when the night falls and covers your spot, you can't see the sun from the new spot and so on.

"Certainly, the non-believers in the unseen will find my quote holding that the night covers the sun is a kind of guessing or prediction of the unseen and therefore, they mockingly assume that it is a kind of sorcery. Likely, the same attitude was prevalent in the age of Messenger of Allah (PBUH)as they used to hurl such insult whenever they saw any God-made miracles, 'we are a people affected by magic' [Al-Hijir:15].

"This attitude was adopted by those who are alike. It is easy for them to believe in a sphere that moves in the galaxy like an atom, orbits around its axis a thousand miles per hour and orbits the sun. This solar system orbits the galaxy as well as the whole galaxy with soaring numbers and they **claim that** it is pure science or proved by science. They are typically of those who would exchange that which is better for that which is lower. They should not be disturbed when being mocked or ridiculed by us as they ridicule us when interpreting cosmic phenomena by the words of Allah, while they believe in the absurdity of modern science, a groundless faith with no substantial evidence. Allah says, '**If you ridicule us, then we will ridicule you just as you ridicule**' [Hood:38]."

25

THE MOON

WHAT ABOUT THE moon? They say that the planets in the sky, the sun, and the moon are all spherical. Then why do we see just one phase of the moon? If the moon is spherical, rotates around its axis, and orbits the Earth, why from time immemorial has nobody seen the other side of the moon except through fabricated space images and movies? This seems weird, but because this is an intriguing query and uproots the very basis of their theories, they came with the idea of "balanced rotation."

The idea is that the moon orbits the Earth as fast as it revolves around its axis so that those who live on Earth can only see one side of the moon. Glory be to Allah! He is All Capable, but is this what really happens? Or is the moon luminous and may be a spherical circular disk spinning in a circular orbit so we can see only one phase of it?

If you consider the idea of the lit disk instead of the lit sphere, you will understand why the face of the moon appears straight at the north pole they look and reserved at the south pole. In reality, both of them are looking at the face of the moon but from different angle, it's not that one is looking at the face of the moon while the other the back side of the moon.

Test it yourself. Hold a quarter of a dirham in your hand and look at the image of the gazelle inscribed on it. Move it as if it were in an orbit on a sheet of paper and draw some points on the paper as human beings. You will note that all human beings cannot look at the back of the quarter, but only the image of the gazelle, but some on one side of the Earth look at half of the gazelle on the other side they look at the other half of the gazelle. None of them can see the other end of the quarter that has 25 fils inscribed on it.

There is another piece of information that I think is important to know, and it is that the light of the moon is not a reflection of the light of the sun, but that it is self-illuminated by the grace of Allah. This is evidenced by the wording of the Quran. Allah, the Almighty, says, **"Blessed is He who has placed in the sky great stars and placed therein a [burning] lamp and luminous moon"** [Al-Forqan:61].

Reflect on the Arabic term "mounira" (luminous). This Arabic word acts as a subject, not object. The moon is self-directed and is the actor with permission from Allah. If one reflects on all the positions in where sunlight or moonlight is mentioned in the holy Quran, one finds that they are expressed in different terms, not the same. When Allah mentions the sun, He, the Almighty, says "Diyaa," and the moon, "Nour." Both indicate illumination but in different ways. **"It is He who made the sun a shining light and the moon a derived light"** [Younis:5].

You will not find the sun is described as "mounira" (luminous) in the Quran.

He also says, **"Blessed is He who has placed in the sky great stars and placed therein a [burning] lamp and luminous moon" [Al-Furqan:61].**

In the chapter of Yūnus, Allah says, **"It is He who made the sun a shining light and the moon a derived light" [Yūnus:5].**

And in the chapter of Noah, Allah says, **"Do you not consider how Allah has created seven heavens in layers" [Noah:15].**

Allah also says in An-Nabaa chapter, **"And made [therein] a burning lamp" [An-Nabaa:13].**

This indicates that the nature of sunlight is different from the nature of the moonlight, and this is obviously sensible as well. Do you not notice the great difference between them?

Weirdly enough, some Christians and others who tend to oppose Islam chose these the verses where the moonlight is mentioned on the grounds that they contradict modern science. It proves that the light of the moon is not a reflection of the sun, and, unfortunately, some of our people replied that their claim is wrong in order to prove that the moon reflects sunlight, and God is Our Assistant!

Both parties fell into the misconception of flatness of earth. The opponents of Islam have the right conclusion but driven by bad intent, whereas others have misunderstood it but with good intent, and guidance is usually in the scale of those who have good faith and the ability to infer correctly, but the Quran does indicate that the moon is luminous, and the Bible also states that the moon has light on its own, not the reflection of sunlight.

We find it is stated in the Book of the Revelation to John, Chapter 6, verses 12, *"(12) And I beheld when he had opened the sixth seal, and, lo, there was a great earthquake; and the sun became black as sackcloth of hair, and the moon became as blood..."*

If the sun becomes black and if the moon reflected its light, why does the moon become like blood while the sun is black? The simple justification offered by some is that this will be known in the future and anything is possible, a technique used by many who do not have satisfactory answers. This does not necessarily mean their view is erratic; it is the best strategy devised to refer to all the Unseen, only known by Allah. What will they say about the following text in the same Book of the Revelation to John, Chapter 22, verse 23, *"And the city had no need of the sun, neither of the moon, to shine in it: for the glory of God did lighten it, and the Lamb is the light thereof."* And, in the book of Isaiah, Chapter 60, verse 19, *"The sun shall be no more thy light by day; neither for brightness shall the moon give light unto thee: but the LORD shall be unto thee an everlasting light, and thy God thy glory."*

I do not know from which source the contemporary Muslims deduce that luminosity means reflection of light. It is apparent that they confused what the commentators said about it being a manifestation with it being a reflector. Even if there is a source of such a view, why should it be accepted, excluding the other views? How can this be done while we do not have any lexicon proving such a view? And from what source do they establish that the moon is dark in itself, while Allah calls it **"luminous"** and **"the moon to be a light**?" If they insist that luminosity means the reflection of light, what will they do with the great Quranic verse that serves as the best refutation of their view, **"Allah is the Light of the heavens and the earth. The example of His light is like** a niche within which is a lamp, the lamp is within glass, the glass as if it were a pearly [white] star lit from [the oil of] a blessed olive tree, neither of the east nor of the west, whose oil would almost glow even if untouched by fire. Light upon light. Allah guides to His light whom He wills. And Allah presents examples for the people, and Allah is Knowing of all things" [An-Nur:35]. Allah is the source of the light, and far be it from Him to be a reflection of something, Exalted is Allah above that, Who said, "And the earth will shine with the light of its Lord" [Az-Zumar:69].

Considering the apparent meanings of the verses and according to good sense, you will not find who would say that the apparent meanings of the verses indicate the moon reflects the light of the sun. Moreover, Ibn Kathīr (may Allah have mercy on him) said when exegeting the verse of the chapter of Al-Furqan, "That is, it is shining and luminous by another light of different type and nature from the light of the sun." As for the contemporaries, due to the fact that they have been influenced by the outputs of the scientific community, they interpreted the verses to have other meanings conforming with such outputs!

Have you ever wondered if the moon is really a reflector of the sunlight? If so, why do we find different colors of the moon all the year round? Also, why is the object under the sun's shadow hotter than the object exposed to moonlight? Moreover, if the moon reflects the sunlight, the said object should then have been affected and get hot, but it does not. It can be said that the sun's shadow covering the object is the cause of heat. That is because the object cannot emit heat stored within it in the open air or a large area, and its temperature thus gets increased. Yet, this answer is not satisfactory, at least for me. When we were young, we used to hold a mirror to reflect the sunbeams, and then we saw the direct impact of such an act. In this regard, for a purpose in my mind, it is better to recall the interpretation of the verse in Al-Furqan, stated by Sheikh Muhammad Al-Amin Al-Shinqiti, the author of the Tafsir book, *Adwaa' al Bayaan*.

"We have previously stated that in the Quranic Chapter of Al-Hijr, the Quran explicitly states that the moon is in the constructed heaven and not in the absolute sky which means whatever is high and above you, because Allah has pointed out in Al-Hijr that the Heaven in which He placed great stars is the Heaven protected and the protected Heaven is the one constructed in His Saying, '**And the heaven We constructed with strength, and indeed, We are [its] expander**' [Adh-Dhariyat:47], and in His Saying, '**And constructed above you seven strong [heavens]**' [An-Naba':12]. It is other than the sky which refers to whatever is high and above you. This is clearly explained in Al-Hijr in His Saying, '**And We have placed within the heaven great stars and have beautified it for the observers. And We have protected it from every devil expelled [from the mercy of Allah]**' [Al-Hijr:16–17]. These verses in Al-Hjir are indicative that the Heaven with great stars is the constructed and protected Heaven rather than the sky that is high and above you. If you learned that, you should know that Allah, Exalted is He, has pointed out in this verse in Al-Furqan that the moon is the Heaven in which the stars are placed because Allah says, '**Blessed is He who has placed in the Heaven great stars and placed therein a [burning] lamp and luminous moon**' [Al-Furqan:61]. This is a proof that it is not the absolute space that is high and above you. The Muslim should not turn away from this apparent meaning except for compelling evidence from what Prophet Muhammad has reported. However, no one should quit the apparent meaning of the Glorious Quran except for a convincing well-known proof.

"Undoubtedly, those who try to ascend to the moon with their machines and claim that they had landed on its surface, their meanness, weakness, and inability will eventually show when confronting with the might of the Creator of the Heavens and the Earth, Exalted is He. We have previously stated with regard to the Quranic Chapter of Al-Hjir that this is indicated by Allah's Saying, '**Or is theirs the dominion of the heavens and the earth and what is between them? Then let them ascend through [any] ways of access. [They are but] soldiers [who will be] defeated there among the companies [of disbelievers]**' [Sad:10–11]. It would be argued that the verses you gave as evidence that the moon is in the protected Heaven may have probably a well-known Arabic style which entails its being not indicative of what it stated, i.e. the pronoun refers to the term only rather than the meaning. That is to say, His Saying, '**Blessed is He who has placed in the Heaven great stars**'

[Al-Furqan:61], refers to the protected Heaven but the adverb 'therein' in His Saying, **'and placed therein a [burning] lamp and luminous moon'** [Al-Furqan:61], refers to the absolute meaning of the term sky (*Sama'*) in terms of being linguistically whatever is above you. This is a well-known Arabic structure expressed by the linguists in saying, 'I have a Dirham and a half,' i.e. half of another Dirham. Similarly, Allah's Saying, **'And no aged person is granted [additional] life nor is his lifespan lessened but that it is in a register'** [Fatir:11], means that the lifespan of another person is not to be lessened. Our reply is that this argument is possible but there is no evidence established to support it and the apparent meaning of the Quran should not be forsaken except for compelling evidence and the apparent meaning of the Quran should be more likely followed rather than the opinions and arguments of the disbelievers and their imitators. And Allah knows better."

In my view, the moon is smaller and closer to us than what the contemporary scientists say. I think its size is the same as the size of the sun or a little smaller than it. Certainly, everything I have stated will contradict their theories and cause a great confusion in the overall system and view they adopt about the cosmos.

If you wonder—and you have the right to do so—about the different forms of the moon or crescents if the moon does not reflect the sunlight, the answer is crystal clear in the Quran, **"And the moon—We have determined for it phases, until it returns [appearing] like the old date stalk"** [Yasin:39].

At-Tabari said, "And this is a sign for them that Allah has fixed stages for the moon where the moon's shape changes from the stage of the full moon to the extent that it becomes in the shape of an old dried date stalk. The Arabic word used in the verse to describe the last stage of the moon is *'Al-'Urjoun,'* which means the dried branch of the curved shape date. Allah, Exalted is He, likens the moon to the old *'Urjoun,'* because the latter is hardly seen but curved and curved when it becomes old and dried and it is hardly to be seen straight and level like the branches of other trees. Similarly, at the end of the month, and before its disappearance, the moon becomes in the shape of an old dried date stalk which is curved and bowed."

Al-Qurtubi and many others said, "The moon goes through 28 stages/mansions (*manazil*) in 28 nights. These lunar mansions are:

"*Ash-Sharaṭayn, Al-Buṭayn, Ath-Thurayyā, Ad-Dabarān, Al-Haq'ah, Al-Han'ah, Adh-Dhirā', An-Nathrah, Aṭ-Ṭarf, Al-Jab'hah, Al-Kharātān, Aṣ-Ṣarfah, Al-'Awwā', As-Simāk, Al-Ghafr, Az-Zubānā, Al-Iklīl, Al-Qalb, Ash-Shawled, An-Na'ā'am, Al-Baldah, Sa'du 'dh-Dhābih, Sa'du 'l-Bul'a, Sa'du 's-Su'ud, Sa'du 'l-'Akhbiyyah, Farghu 'd-Dalū 'l-Muqdim, Farghu 'd-Dalū 'l-Mu'khar,* and *Buṭnu 'l-Ḥūt.*

"When the moon reaches the last of them, it goes back to the first one. In its orbit, the moon moves through these mansions in twenty-eight nights, then it disappears and then emerges in the shape of a crescent. Then it returns to moving in its orbit among the mansions which are associated with a dominant star or constellation. Each constellation is linked to two mansions and a third. For example, Aries (Burju 'l-Ḥamal) is associated with *Ash-Sharaṭayn, Al-Buṭayn,* and the third of *Ath-Thurayyā,* and Taurus (Burju 'th-Thūr) is associated with the two thirds of *Ath-Thurayyā, Ad-Dabarān* and two thirds of *Al-Haq'ah,* and so on."

These lunar mansions are the cause of changing the shape of the moon, and its shape changes when it reaches each one of them. Let us consider a simple example. Have you ever gone to an amusement park—or have you taken children there—and visited a room with a large number of funny lenses or mirrors? Sometimes, when you are in front of a mirror and look at the reflection, you would see yourself bigger or smaller, taller or shorter, or your face may seem concave, or a part of your body would look larger, and so forth. In my opinion, this is similar to what occurs to the moon. The moon has mansions, and as soon as it reaches one of them as a guest, the shape of the moon that appears to us gets changed, just as what happens with the filters on mobile phones, and Allah knows best.

26

SOLAR AND LUNAR ECLIPSES

H ow many times did we see the sun and moon together at the time of solar and lunar eclipse? I saw that, as did many other people. This is a sufficient evidence that the scientists' theory is not correct, because it holds that the moon—during the lunar eclipse—lies on one side while the sun lies on the other side, and the Earth stands between them. Then, how does the lunar eclipse happen as it is the appearance of the shadow of the earth reflected on the face of the moon?

Pay attention that when the moon is on one side, the sun on the other side, and the Earth is between them at the time of lunar eclipse. This means the three of them are not in a straight line. Given the supposed motion of the Earth around its axis, the movement of the moon around its axis, its similar movement around the Earth, the movement of both the Earth and the moon around the sun at magnificent speed, the movement of the sun inside the constellation, and the movement of the constellation inside this universe in momentous celestial numbers, it would be strange to see lunar eclipses taking place regularly and be easily defined.

Also, let us consider why we do not see the lunar eclipse on Venus or Mercury. This is because such planets have no moon like ours, and even if they had, their nature would be different. If they had a moon similar to the Earth's moon, the distance between the moon and the planet would be different. Even if the distance was the same, our sight has limited abilities. Allah, Almighty, says,

"They would say, 'Our eyes have only been dazzled. Rather, we are a people affected by magic'" [Al Hijr:15]. These scientists would believe in anything other than the easy and simple explanation held by both commoners and scholars. These scientists like to believe only in scientists' words—those whom they consider to be scientists. It follows that scientists who hold views unacceptable to them are deprived of the title of a scientist. In Allah we seek help!

Why is the shadow smaller than the moon while the distance between the Earth and the moon is known to the scientific community? The shadow should expand and get bigger, farther the moon is from the Earth and vice versa. Since the moon is 385 lm from the Earth and 147 million km from the sun, and the Earth is 150 million km from the sun, how can the alleged earth shadow on the moon conform to these dimensions? And if what we see reflected on the moon's face is truly the Earth's shadow, why does it have different colors? The moon also has different colors during the lunar eclipse, and the color of the shadow is known. Similarly, why have people throughout history seen both the sun and the moon on the horizon despite it being the time of lunar eclipse?

If you believe in the theory of the spherical shape of earth and then discovered that there are people inhabiting distant areas of the earth and see the same thing, you would certainly wonder how this could possibly happen. If the reason for the occurrence of lunar eclipses is the earth's shadow, why don't we see the lunar eclipse when the moon faces the center of the earth in the center of the sky and where exactly the center of both is from these scientists' perspective in the first place?

By Allah, it is more than strange that some modernists reproach and those who imitate Muslims and hold them inferior while they do not reproach themselves or consider themselves inferior for following theories unsubstantiated with irrefutable indisputable logical evidence or even a single Quranic incontrovertible evidence. In Allah Alone we seek help!

Anyway, in the chapter on Gravity, we will discuss now the phenomenon of lunar eclipse undermines these scientists' theories. If you said that today's scientists can predict the times of the occurrence of lunar eclipses, I would retort that scientists could do that in pre-Islamic ancient civilizations, even before Christ, using the Saros Cycle, or even with the astronomical tables that were designed to facilitate the calculation planetary positions, lunar phases, eclipses, and calendrical information, and so on. The fact that this is done nowadays because of scientists' modern equipment does not mean scientists in the past could not do it. It also does not mean those advocating the theory of a flat earth cannot determine when lunar eclipses happen.

A man with an artificial arm can make food no better than that of natural arms. Though an artificial arm can be made of iron and can have techniques and capabilities greater than the natural one, it does not mean that a man with a natural arm is incapable of making his own food. Nowadays, they just need to define the positions of both the sun and the moon, make some computerized calculations, come up with some scenarios, and it is ready. The relevant codes are available on the internet, and those interested can use them and predict the occurrences of lunar eclipse.

I once read in a book—whose title unfortunately skipped my mind—that astrologers gave Alexander, the Great, a chart containing their calculations of the lunar eclipses for the then upcoming two thousand years. Allah only knows if this really happened. However, this shows that the idea of speculating the occurrence of solar and lunar eclipses is too old and existed centuries before the prevalence of the spherical Earth theory.

When it comes to the lunar eclipse, it makes no difference if you believe in either the spherical or flat shape of Earth, once you think that the sun and the moon are almost of the same size, that they are very close to the earth, and that they are way smaller than modern science states. It is easy to explain the occurrence of lunar eclipse on the basis of a flat Earth. You just say that the sun and the Earth as described by Allah Almighty, **"all [heavenly bodies] in an orbit are swimming"** [Al Anbiyaa:33] move in a circular orbit, and the solar eclipse is caused by the moon.

It can also be said that the sun and the moon are both transparent bodies. When they are both within the range of sight, people see the lunar eclipse. You may also say that they both move in their orbits and when they meet in a point, the solar eclipse happens.

There are several interesting hypotheses that entice man to more research and can apply to both the theories of spherical and flat earth, but do I have an unquestionable opinion? In fact—and to be honest with you and with myself—I do not have one yet. This is because I still have this sight of limited nature. Allah Almighty says, **"It is not allowable for the sun to reach the moon, nor does the night overtake the day, but each, in an orbit, is swimming"** [Yasin:40]. Some of the hypotheses on solar eclipse attribute it to the fact that the sun catches up with the moon. Yet, I hold that the sun revolves in one orbit different from that of the moon and when they become parallel and in the same range of sight, the solar eclipse happens, and Allah knows best.

Moreover, you may believe in the idea that there are transparent or invisible celestial bodies causing the phenomenon. You may also hold that the source of the light of both the sun and the moon is in the celestial sphere, and we see only the bodies reflecting this light. Thus, when some bodies block the way between the source and the reflective body, the phenomenon occurs. It can also be said that the solar eclipse is attributed to a characteristic peculiar to the sun, and that the sun causes its own lunar eclipse independently from the moon. So far, I am not fully aware of the details of such hypotheses. I wish you can come up with a hypothesis inspired by reading this book and studying the issue. Once you come up with an answer, please let me be informed and write to me, and I would be grateful for that.

In short, there are several answers and ideas whose detailed pieces of evidence need careful study. They are available to all. One of them may be the right answer, and all are relevant to the theory of flat Earth. All praise is due to Allah, Lord of the worlds!

Finally, I would like to note that the phenomenon of the giant moon or the super moon should raise concern for those who advocate the spherical shape of the Earth or gravity. Why at the time of a giant moon which is closer to Earth do we notice no kind of changes on the tide and ebb on Earth and the like? What makes the moon draw near to us? Is it gravity or some other reason? What about the moon's moving away from the Earth once more? What gave it the strength required to resist the Earth's gravity to draw away from it and regain its normal appearance? This is strange, isn't it? It is true that people have linked catastrophes and disasters with the phenomenon of giant moon or super moon, but even scientists of modern times deny that and consider such claims phony.

27

DELUGE OF NOAH

IT IS WELL known that Allah drowned the world by the deluge of Prophet Noah (peace be upon him). Yet, recently, some different opinions in this regard have emerged, as some said it was not a worldwide flood but only for the people of Noah. In fact, I have reviewed some opinions on Noah's deluge in detail in the third part of my novel, *7000 Worlds*, but we could briefly touch upon some aspects of them. Earlier, I used to be one of those who firm believed that the deluge was worldwide, but after reading the books of the contemporaries and listening to their lectures on the deluge, I questioned the matter, and said, "Perhaps it was not worldwide and did not cover the whole earth." However, in the end, I have become reassured in my heart that the deluge was indeed worldwide.

Allah has mentioned Prophet Noah more than forty times in the Noble Quran and did not state in any verse that he was sent to the world. Allah, the Almighty said, **"Whoever is guided is only guided for [the benefit of] his soul. And whoever errs only errs against it. And no bearer of burdens will bear the burden of another. And never would We punish until We sent a messenger"** [Al-Israa:15]. Therefore, if Prophet Noah was not sent to the world, it would be impossible that Allah punished the whole world. What did happen, then? Allah, Exalted is He, said, **"And it sailed with them through waves like mountains"** [Hud:43]. Besides, it did not avail Noah's son that he took refuge on a mountain to protect himself from the water, as stated in the Quran, "[But] he said, 'I will take refuge on a mountain to protect me from the water.' [Noah] said, **'There is no protector today from the decree of Allah, except for whom He gives mercy.' And the waves came between them, and he was among the drowned"** [Hud:43]. Thus, if the deluge was not worldwide, how could water reach the mountaintops when we live, as they say, on a ball floating in the vast space? In addition, why did Allah command Prophet Noah to carry two animals of each kind—male and female—in the ark? **"[So it was], until when Our command came and the oven overflowed, We said, 'Load upon the ship of each [creature] two mates'"** [Hud:40]. If the deluge was not worldwide, then what caused Prophet Noah to take the animals, when he could have brought them from another part of the earth? What exactly did happen? Interestingly, the apparent meanings of the Torah's texts indicate the deluge was worldwide, but that is not explicitly stated in the Quran. Allah did not say even for once in the Quran in the story of Prophet Noah that He drowned all of the Earth. Instead, He said: **"But they denied him, so We saved him and those who were with him in the ship. And We drowned those who denied Our signs. Indeed, they were a blind people"** [Al-a'rāf:64]. Is it possible that the flood was worldwide? If it was so, how come Allah would drown the people to whom Noah's call to Allah did not reach? Or were the people of Noah the only humankind living on Earth at the time? Weren't there any other people on Earth at the time? What about animals? Were there no other animals on Earth at the time other than the animals in the region where Noah was living? Additionally, Allah said, **"And the people of Noah—when they denied the messengers, We drowned them, and We made them for mankind a sign. And We have prepared for the wrongdoers a painful punishment"** [Al-Furqan:37]. People is a general word that may refer to the people who coexisted with Noah, people who were living in cities, villages and other continents other than Noah, those who would come after the people of Noah, those who were with Noah and survived from the deluge, or may refer to others . . . It is very confusing.

There must be an answer for these questions, and I cannot really imagine how water could reach the mountain peaks, in the perspective of the spherical

Earth, only with rain, waves and tsunamis and the like. Besides, this does not happen in reality, as water seeks a flat surface to settle on. If the Earth is spherical, the water will seek a flat surface to settle on. However, if it is flat, drowning the world will be logical and comprehensible, maybe not by the average rainfall or rough waves, but as Allah said, **"(11) Then We opened the gates of the heaven with rain pouring down. (12) And caused the earth to burst with springs, and the waters met for a matter already predestined."** [Al-Qamar:11–12]. This is apart from what I quoted to you previously from the Torah in this regard, *"all the fountains of the great deep burst open and the floodgates of the heavens were opened."* If you imagine yourself living on a flat Earth and the heavens cover you like a ceiling with water present over the heaven and in the depth of the Earth, then the world can become fully filled with water so that the water reaches the mountaintops. The matter is thus reasonable. It is known that when the rain comes down from clouds, it falls in drops. However, it should be noted that Allah said in the said verse, "pouring down." If you store water using a wall, window, or the like, what will happen when you remove the barrier? The water will pour down. If you remember that it is stated in the Torah that God separated the "waters above" the Earth from the "waters below" the Earth by a firmament or Vault, which is the heaven or the dome. Thus, if the gates of the heavenly dome are opened, what would be expected to occur? The water would pour down. Glory is to Allah, there is a wonderful agreement between the Quran and the Torah in this respect.

Noah's story is mentioned in a number of religions and myths. In Hinduism, we find the story of Manu, and it is stated in some narration of this story that god Brahma flooded the whole universe, not just the Earth. Even the latter point, which is considered inconceivable, would become somewhat logical if we consider the hypothesis that the whole heavens and Earth altogether form the cosmos according to our contemporary concept of the cosmos, and that we live within this giant cosmic bubble. Therefore, it would be normal that the universe gets drowned if such bubble or closed cage is filled with water. In short, if you accept the spherical Earth perspective, then the fact that the water reached the mountaintops will be problematic, aside from the deluge being worldwide and flooding the whole world. However, if you consider the flat Earth perspective and that the heaven is like a dome, then any narration of the deluge can be interpreted by a logical interpretation that is consistent with the flat perspective, and Allah, the Almighty, knows best.

28

GRAVITY

G RAVITY. IT IS the super force that holds millions of tons of oceans but is incapable of pulling birds down. It is the great power that makes the moon revolve regularly in our orbit but is incapable of pulling down insects. It is the great force that makes fruits fall down but is incapable of holding the petals of a weak rose from floating in the air. There is no doubt that it is a rational force, for it can pick and choose the items it attracts and the things on which it does not tighten its grip. The status of gravity in cosmology is like the domino piece that has been a foundation for a great house of domino pieces. Once you draw the piece of domino that holds up the whole building, the great building you took pains to establish, would fall apart. If the theory of the gravity of the Earth and the sun was refuted, the whole building of modern science of the Earth, the sun, and the moon and other important elements of cosmology would crumble.

Is gravity a fact or a theory? In fact, it is a theory. How was Newton inspired to put his theory in the first place? By the falling down of an apple. Isn't this true? At the time, he thought gravity is the reason behind the falling down of the apple. Well, Newton, it is okay. But why didn't the rest of the fruits fall down as well? Why did this fruit in particular fall down? Did the falling apple stand still on the ground, or did it roll on it? If it rolled, why did this happen while gravity was so strong that it forced the apple to fall to the ground? It is okay, Newton. However, did only one apple fall, or did all the apples on the tree fall? If only one apple fell down, why was it this one in particular? Isn't this because heavy bodies fall and need something to support them and prevent them from falling? When the fruit falls down, this is because the branch holding it has grown weak and the bond between the fruit and the branch gets weak when the branch (the link) grows incapable of carrying the weight (the fruit's weight). As a result, the fruit is separated from the branch and when this takes place, the fruit needs a source of support that prevents it from falling? What is the first source of support it gets? It is air. However, since air cannot bear the fruit—for its density and weight is less than the density and weight of the fruit—the apple falls down. Then, it needs to settle down, and it finds settlement in the ground. This is the nature of the things that have weight, and it is known to everyone. So, why should it be called gravity, now?

Let us presume that Newton's theory that the fruit fell down because of the Earth's gravity was a correct one, why should the Earth's gravity be a proof of the existence of gravity in all bodies, whether the sun, the moon, or the like? What can guarantee that the laws here on the Earth are the same as the laws on the sun and the moon? Did anyone actually stand on the sun to witness an incident similar to the one Newton did on Earth, i.e., a fruit falling to the ground? It is said that the Earth is a dark body and that the sun provides it with light. Yet, if we were to follow their way of thinking, we would have said that the sun must be a dark body as well, because the Earth is a dark body. As a consequence, there must be another luminous body that provides the sun with light and so on. Do you see the conundrum they face because of generalizing upon the theory of gravity and using it as an explanation? Generalization is not recommended in such matters, especially when man had not discovered a lot about celestial bodies.

If I agree that this theory can explain the incident, I can also say that I can explain everything by divination. I can well say that on falling, the fruit was ordered by the angel to whom it is assigned to fall in this specific spot. Similarly, the angel assigned with the sun is the reason why it attracts all the other bodies, and he steers them and makes them revolve around its orbit. Now, can you prove me wrong? You cannot, because you cannot put my theory to the test, for you cannot see the angel, nor can you see gravity or put it to the test. Explaining things using an idea is something, and proving it is a completely different matter. In fact, I do not understand when people can explain a phenomenon or a given incident with a tangible evidence that can be seen, experimented, and acknowledged, but they choose to ignore it and go for another idea or theory that cannot be proven right or put to the test, just because they simply believe in it. I can totally relate when they do so in tackling extraordinary matters or metaphysical things, but the case is different when it comes to science and scientific experiments. Unfortunately, their method is fluctuating according to their mood. It follows their hearts and not facts.

Anyway, when Newton put his theory, Einstein shattered it with his theory of space-time, though Einstein's theory is not beyond doubt either. Up till this day, if you did a humble research effort, you would discover that there are issues and challenges facing the gravity theory, and that scientists came up with new theories, some of which correct the theory to improve upon it and others refute it altogether. Thus, I wonder why the meanings of the Holy Quran are interpreted in a way that deviates from their apparent significance to conform to a theory that can be either accepted or refuted. Doesn't the rule of interpreting the Holy verses stipulate that in the absence of irrefutable evidence that this is not the case, they should be interpreted at face value? And where are the experiments that prove the gravity theory?

It is said that there is an experiment we can conduct. If you put an apple and a feather in a closed place void of air, the falling velocity of the apple would be equal to that of the feather. Can you imagine that? But the question is where would we find such a room void of air? And what would happen in a space void of air? It would be squeezed, so how could there be a space?

Unfortunately, the media manipulated the public through conducting this experiment to prove the existence of gravity and the public believed the experiment to be true. In the experiment, the feather and the apple fell simultaneously. However, the viewers did not realize several things. When the apple and the feather fell to the ground, they bounced back. How could this happen in the absence of air? If there was no air, gravity should have pulled them down, and no opposite force, like air, would be there to cause them to rebound. Thus, how did the bodies bounce back? Moreover, the feather bounced back stronger than the apple did, meaning that it bounced back further than the apple. Why did this happen in a room void of air?

They also proclaim that the higher you go, the lighter weights get. So, presume that technology achieved fascinating advances, and we became capable of cutting the mountains and taking them along with us remotely in the space! According to them, if this happened, the weight of the mountain would decrease. Isn't it right? Thus, if we reached a point where the mountain's weight was equal to a feather's weight and one of the astronauts was provided with the equipment necessary to enable him to walk on a planet invaded by us in the future, according to their hypotheses, that man would be able to lift the mountain with a single finger of his. In the Hadith attributed to Prophet Muhammad (peace and blessings of Allah be upon him), according to 'Abdullah Ibn Mas'ood, may Allah be pleased with him, a Jew came to the Prophet (peace and blessings of Allah be upon him) and said, "O Muhammad! Allah holds the heavens on one finger, the Earth on another finger, the mountains on another finger, the trees on another finger, and the created beings on another finger, then He says, 'I am the King.' The Messenger of Allah (peace and blessings of Allah be upon him) smiled so broadly that his eyeteeth could be seen, then he recited, **'They have not appraised Allah with true appraisal'** [Az-zumor:67]." Another version of the Hadith reads, "Allah's Messenger smiled in wonder and affirmation of what he said." Regardless of how sound the Hadith is, imagine there is a man who can lift a mountain with a single finger before people. On reading in the Holy Quran that Allah moves the mountains or blows the mountains away with a blast on Doomsday, would that man feel the magnitude of such actions, or would he rather say, "I can do this here in this worldly life and can hold the mountain upside down to play with it as if it was a kid playing the ball?" Which would be the case then, Only Allah knows.

Another feature of gravity is that without it, bodies have no weight. Gravity gives the bodies their weight according to the type of the matter, its shape, its volume, and the like. For those advocating gravity, if it weren't for gravity, the mountain would have been the same weight of a mustard seed. If this was the case, the measuring scales would be of no value. People would not be able to rely on weights because they would need to measure the force of air resistance as well. In short, the weighing systems would be disturbed if we agree with them on that. It is certain that gravity cannot explain why huge steel balls fall rapidly; unripe fruits remain on the tree without falling while Newton's apple fell down.

How were the continents formed? And what about the theory of the continental movement or what is called the continental drift? If the center of gravity is at the center of the Earth, continents would have never moved apart the way we see today. How do the plate tectonics occur in the first place? Think deeply about it, and you will understand why I mentioned it.

Let us take two stones, one of which is way heavier than the other. Which of them would fall first? The big one would, right? The question is why? If gravity and centrifugal forces differ according to volumes, shouldn't the force of gravity be of greater influence on the small stone, given its resistance is less than the other big one? Any answer the theory's advocates give would lead to conundrums and questions they should answer, but how could they?

Here is another issue. If the theory advocates say the Earth's gravity force is of fixed value and does not change, then fetch a cup. Fill half of it with water and the other half with oil. What would happen? The oil would float. Why? If gravity force had a fixed value, why would this happen? Throw a log and a pebble into the water. The log would float. Why? If the influence of the Earth's gravitational pull is the same, why would this happen? These experiments we used to conduct in our childhood are important in proving the flaws in the theory of gravity in which modern scientific societies believe. The same applies to vapor and smoke. Why does smoke go upwards in the air? If your answer is, "The air pushes it upwards," don't the theory advocates say that gravity pull holds the Earth atmosphere down so that man never senses that the Earth moves? And what are the vapor molecules?

In the same manner, if the Earth revolves around the sun because of the sun's gravity, why don't we ever feel the gravity of the sun? If we travel in a balloon, fly on a plane, or launch rockets, why do none of us feel the sun's gravity affect us? Isn't it true that the sun's gravity affects the Earth? What about the moon's gravity? There are many questions that those who advocate the theory that the Earth is spherical would find difficult to answer, the same way scholars find it difficult to make a weak Hadith conform to a Quranic verse with an eye to defending a weak Hadith deemed to be sound.

The advocates of this theory said that the sun can accommodate more than a million and three hundred thousand Earths inside it, due to its humongous size, and they proclaim the Earth came to existence more than 4.5 billion years ago. No problem. Let me assume that this is what actually happened. If this is true, why didn't the sun pull the Earth till it collided with it? Isn't the sun's gravity and size much bigger than that of the Earth? They state that the weight of the sun is 333,000 times the weight of the Earth. Then why wasn't the Earth pulled toward the sun to collide with it more than 4.5 billion years ago? Is the Earth's centrifugal force strong so that it has not collided with the sun till now? Imagine that a man pulls a rope against 333,000 men, given that all the men are equally strong. Is it possible that 333,000 men are trying to pull one single man on the other side of the rope for more than 40 years and fail to do so?

Having talked about the sun, I would now talk about the moon. There is a concept widely accepted in the scientific circles known as the Big Bang theory. Briefly, it tells us that the moon came into existence 4.5 billion years ago. Imagine that huge number. Suppose the moon came into existence even just a million years ago, what makes the moon float in its orbit above us since man came to existence up till now? Why didn't the moon draw away from us and closer to the sun? Doesn't gravity have the final word? It is true that gravity pull is affected by the mass and distance. Then, have you ever asked about the size of the sun in comparison to that of the moon? The moon is way smaller than the Earth. You are already familiar with the size of the sun in comparison with that of the Earth. Thus, why didn't the sun pull the moon? Even if we said that the sun is way farther than the moon, the size of the sun would be of a strong influence in spite of the distance between the moon and the sun. The moon should have been drawn closer to the sun, but this did not happen. If the answer is that the moon draws away from us gradually, I will retort, "Why don't we feel that?" The moon is the same as described thousands of years ago, in the same positions and dimensions. Why didn't anyone notice it? If I agreed with the assumption that the moon gets farther and farther, if this is the case, the moon should have collided with or dissolved in the sun hundreds of thousands of years ago. When the moon gets closer to the sun, the force of the sun's gravity gets stronger, and consequently, the moon would draw closer to it more rapidly. However, this didn't happen. It seems that the moon does its best to stay with the Earth.

It is said that the lunar eclipse occurs when the Earth shadow screens the sunlight reflected on the moon. According to them, this happens when the three bodies—the sun, the moon, and the Earth—are in a state of alignment. Isn't it so? Now, this hypothesis refutes the theory of gravity. If you ask, "How?" the answer would be as follows. The sun has its own gravity, and so does the Earth. Since the eclipse happens when the Earth is positioned between the moon and the sun, this means the sun and the Earth lie on one side and the moon lies on the other side. Since the moon is on the other side, it is subjected to a stronger gravity pull. Both the sun and the Earth pull the moon toward them, and they lie on the same side. Thus, the moon should have been pulled toward the Earth and collided with it or drawn closer to the sun because of the eclipse. Yet, this doesn't happen. The moon keeps floating in its orbit as it did since man watched its movements thousands of years before Christ and till today. This means the lunar eclipse happened at least 5,000 times. Yet, the moon did not draw closer to the Earth, and history books never mentioned that the moon collided with either the Earth or the sun. Since the advocates of the gravity theory say that the moon came to exist more than 4.5 billion years ago, you can imagine how many moon eclipses took place ever since. The number would be for sure countless.

Do you want to try it yourself? Buy a small car toy a toddler can play with. Tie it with a strong string and tie the other side of the rope in your hand and try to pull the car toward you. What would happen? The car toy would be draw toward you. Isn't this true? Imagine there is a wrestler a hundred times stronger than you, standing behind you, and pulling you toward him. Would the toy stay in its place where you stand, or would it be pulled nearer to both you and the wrestler? It would certainly be drawn to you. This is the case with the moon and the sun. Strangely enough, the moon resists the gravity forces of both the Earth and the sun combined when they are on the same side. Since the moon is neither pulled toward the Earth or the sun, they can well say that an ant can resist the pulling force of a huge animal.

The same applies to the movement of the planets and the Earth's position compared to their positions. With the change of the planets positions, the velocity of the Earth should change because of the influence of gravity, whether on the Earth or the solar system planets. Yet, this does not happen. This proves the Earth is not affected by the increase or decrease of the distance between it and the other planets. This shows us that the theory of gravity is incapable of explaining this and, thus, what I have mentioned refutes the theory that the Earth revolves and moves and proves that it is rather a stable continent.

Unfortunately, nowadays, the theory advocates cannot come with an explanation that does not go counter to their theories. Contemplate all this to know the great challenges the gravitation theory faces. Their theories refute one another. Once they discover one problem in these theories, they attempt to solve this through apologies and coming up with answers. Yet, once they do, other problems erupt. The case is not so with the Holy Quran. Once the Quran faces any doubts, these doubts turn out to be a new evidence that proves the greatness of the Holy Quran and renders the believer's belief stronger in his Lord's Book and finds himself reciting the words of Allah Almighty, **"[All] praise is [due] to Allah, who has sent down upon His Servant the Book and has not made therein any deviance"** [Al-Kahf:1].

To make a long story short, when the density and mass of the bodies are heavier than the density of air or when the bodies molecules are heavier than the air molecules, they fall. If the body's density and mass are too heavy for air to carry, it falls, else it simply flies. To prove that, all you have to do is hold one of the balls and let go of it. What would happen then? The ball would fall to the ground. They called this gravity. Right? Well, hold the same ball and go to any pool nearby, or even to the beach. Try hard to put this ball under the water so that it goes to the depth of the sea. What would happen? Yes, the ball floats however hard you try to keep it under the water. Isn't it true? In short, this is because the density of the surrounding water is equal or greater than the ball's density. Al Fakhr Al-Razy summed the answer up in the following words, "What's heavier than water sinks into water and never floats onto it."

It might be said, "This does not explain why the sun, the moon, and the stars do not fall. Why don't they fall?" The answer, my dear reader, is that you should know that there is "up" and "down." Things by their nature fall down except for those that are already down like the ground that is the bottom line of the world where we live. That's why Allah holds the heavens so as not to fall on the ground and holds the birds aloft in the sky. Anything is bound to fall after depleting its anti-falling energy. If you asked: "What about the sun, the moon and the stars," I would say, "The sun and the moon float in orbits and the orbits are in the sky in the shape of rings or paths. If you said, "What about the stars," I'd say: "Because they are hung from the celestial sphere like lanterns." And if you said, "What about the clouds," I would say "Because Allah subjugated it to exist between the sky and the Earth, this is one of His signs.

Finally, it is important to say that I did not deny the existence of the gravity force of the sun, the moon or the Earth. I do not deny the existence of gravity, in general. We all know that the magnet has gravitational force. However, the magnet's ability to attract the bodies does not mean each and every body in the world has the same power to attract. Similarly, the fact that water as a liquid has a permanently flat surface does not mean all the matters in the universe are as such.

As for the magnet, have you ever used it one day? You must have used it in your childhood as I did. When matter gets drawn to the magnet, have you ever wondered, "Is the magnet's gravity greater than that of the Earth?" This question alone may bring about a change. Things may be uncovered and other facts could be unraveled to you, Allah Alone is the Guide.

29

SATELLITES

I HAVE ENGAGED IN discussions with advocates of the theory of the spherical shape of Earth. These usually end up with persuading them that the Earth is flat. Yet, the recurrent problem is the existence of satellites. How can the Earth be flat with the existence of hundreds of satellites? And what do you think of the International Space Station (ISS)? What they do not realize is that whatever names they assign to the space station, it would not change the fact that it is not in space. You can well call a cat "lion," but this name would not change the fact that it is a cat. Isn't this true? Satellites, spaceships, and the astronauts' visits to them all go counter to the fact that the Earth is flat. You may ask, "How is this possible?"

The answer is simply that these satellites and stations, though existent, did not go beyond the sky. They only float in the atmosphere. You have read previously in this book in the chapter entitled *The Chapter of Heavens and Earth* that the sky is a protected ceiling and that men and birds are allowed to fly only in the surrounding atmosphere, that is, the distance that lies between the sky ceiling and the Earth. Once you understand that, you would learn that there is no problem in having satellites or space stations. As planes fly in the sky, so do the satellites. They do not go beyond the sky.

But, if they were floating in the atmosphere, there is a problem. Can you imagine what it is? It is gravity. Yes, gravity should have pulled the floating satellites down. The satellites resist gravity, and to keep performing their assigned mission, they need power. However, what kind of engines provide the satellites the power necessary for flying for such long times in the sky?

If they said they float in space, this can be refuted in so many ways using their own theories such as gravity, the revolving of the Earth around the sun and around its axis, the moon, penetrating Van Allen's belt, and many other theories.

It suffices to ask yourself, "How can a satellite revolve around the Earth with the same speed of the revolving earth in order to keep working? They may say that they put satellites within the Earth's gravity field and then leave the rest to gravity, but what would they say about the moon that draws closer and then away from the earth? Does it not have its own gravity that causes the tides? Even if we agreed to this assumption, what is the power that enables it to work that long? Why aren't the receivers in continuous movement, especially with their assumption that the Earth moves in a circular way around the sun? Since the Earth moves that way, what is the power that enables the satellites to work, since either the satellites or the receivers must move?

As for the statements of the (ISS), tell me about it, especially the photos of the spherical Earth. This is unacceptable. As aforementioned, these photos are fabricated. They are not real. Nowadays, we do not have a single real photo of the whole Earth. By this, I mean a live broadcast photo. The live broadcast is always transmitting photos of parts of the Earth and not the whole Earth. You can make sure of this by referring to all the live photos of Earth on their official sites and, in my opinion, it is improbable that anything that floats in space can photograph the whole Earth. This is because the Earth is much greater than we can imagine, and it is bigger than its own atmosphere. Further, it has high, towering ends, according to the pieces of evidence I collected. I mentioned some of these within this book and abstained from mentioning others.

Have you ever conducted research regarding the International Space Station? On Wikipedia, you will find ample information that will amaze you and entice you to pose some questions. The space station was launched in 1998, more than twenty years ago. It weighs more than 400,000 kg and is 70 m in length and more than 100 m in width. Its speed exceeds 25,000 km per hour, and it revolves around the Earth once in every 92 minutes, 15 cycles a day.

Till the present day, it revolved more than 100,000 cycles around the Earth. Imagine having all this momentous energy, but still not taking a single image of the whole Earth. We don't even have one single live photo of the International Space Station revolving around the Earth. Imagine that! They possess the gigantic capability of making this station with the specifications I mentioned and replacing its passengers, even though it is revolving that fast, but they are unable to take a live photo of the space station where it is seen floating around the whole Earth sphere in its orbit.

If asked whether I believe in the existence of satellites, I would say that I do doubt their existence. This is because I have never found a single real photo of satellites. If they really exist in hundreds, we should have found several photos of them.

How can they photograph distant planets and invent technology that photographs the earthquakes on Mars and the like? I do not want them to go so far. All I want from them is one real photo of these satellites to clearly see the space station. They do not have such a photo, today, but I do not exclude the possibility that they would show people in the next years many photos using virtual reality or augmented reality techniques and other modern techniques. We often hear the statement that space investments and space sector are vital. Well, why don't you make an exhibition of whole satellites, not parts of it? Show us the final phase satellite that is ready for launch. Why don't you make such an exhibition? I never attended such exhibitions. I do not want to see just three-dimensional shapes, but true equipment. If this is not possible, why don't we see live images of all the satellites or live photos of them floating in the space?

Further, why are all the receiver dishes that are on the houses not directed toward the sky? They do take slightly require a horizontal orientation. You will find that most of them even are facing the same direction. Why? If satellites revolve around the Earth, why does not the direction of these receivers change? And if the land is of a spherical shape, the direction of these dishes should change so as to choose the direction closest to the satellite, if the whole thing was not but a mere simulation and development of "Loran" navigation system.

Finally, I would like to say that those who launched satellites should answer these questions, and they have to come up with substantiating evidence. The burden of proof always rests on the plaintiff. At the end of the day, if these satellites or space stations exist for real, they would not be in the space; they would rather be floating in the sky. They would not be revolving around the Earth but rather over the Earth. This neither goes counter to the flatness of the earth nor evidences that the Earth is spherical. The flatness of the Earth would be more of a correct idea because of the many problems that would erupt regarding these satellites and the station if we assumed that the Earth is spherical and merged it with other theories such as gravity.

30

MOON LANDING

THEY CLAIM NEIL Armstrong was the first person to set foot on the moon in 1969, that is, about fifty years ago. I wonder, if the invasion of space, construction of colonies, and observation of planets give super advantages, why has the United States not visited the moon after their first visit, and why they don't go and bring any of the moon's rocks or soil or any object thereon to sell to countries or put in museums?

The technology has reached a greater level of development than that at the time of the moon landing, so they should have been able to land on the moon annually. If you say they lost that technology, as some of them said, I will then say that they have the technology that has enabled you to reach Mars or close to it, but at the same time, you cannot reach the moon? As for the rocks brought back from the moon, as they claimed, and which were given as a gift to the leaders of the countries, many of them were vanished or stolen! Even the rock Armstrong gave to the Dutch prime minister at that time turned out to be just a piece of petrified wood! Not only that, the moon landing fact tapes of Apollo 11 are missing. In 2006, NASA admitted that it had lost all of the moon landing records. In 2009, newspapers published that NASA acknowledged the said matter and was working with Lowry Digital, a company specializing in the reproduction of classic Hollywood films, to technically reproduce the moon landing, relying on some recordings in the archive that are not considered live ones.

Consider their statements and admissions and how they conflict and contradict with each other.

In this regard, I recall I was once present at one of the government summits and Michio Kaku, who is considered one of the contemporary greatest scientists in the field of theoretical physics, was there. He said that day, "The technology applied today in the smart phones is more advanced than the technology that the astronauts who worked on the moon landing had." I have no doubt about that, but if you actually landed on the moon, and if what you, the great theoretical physicist, say is true, then why do not you land on the moon once or twice every year? While I was watching the audience clap for his speech without asking any question, I was really saddened by the lack of collective awareness.

I also remember that after he finished his lecture, and during the break, I spoke to one of the attendants. I said to him, "Michio's speech seems exciting, right? But I wonder why the technology is not utilized to land on the moon again, and why there are no mementos brought from the moon?" I then felt that as if a cold water was poured on the person who I was talking to and as if he suddenly woke up from deep sleep. I do not know what happened with him after that, but I hope that my question motivated him to search instead of keeping emulating blindly.

You may wonder why they would have lied about landing on the moon. There could be many reasons. The US President might then have believed that the moon was spherical with an exact Earth-like body on which they could. He might have been a man of vision and wanted to put his people and country ahead, and that is why He ordered the specialists to land on the moon for those national purposes. And, since it is difficult to disobey the president, they found no way out but to deceive the President and claimed the fabricated the moon landing, thus achieving the desired national propaganda. That is how it ended up, just as what occurs in some Gulf countries when management consultants are hired. They do not do anything, but they often take the ideas of the employees and present them in a new way. That is why they could not repeat the moon landing after that, and as one of their employees said they lost the technology.

The reason might also have been a challenge to the Soviet Union at the time, as they were in the Cold War. Or it might be a publicity stunt, just as many governments do when they take reprehensible actions, then overhype and highlight certain positive events or phenomena to attract public's attention away from monitoring their awful acts. There may be many other reasons. The point here is that the reasons for lying and wrongdoing are countless.

There many documentary films spreading on the internet that prove the human landing on the moon is a mere lie. Also, the videos and articles of researchers in this field are much more than the documentary films, most of which are not devoid of realistic and logical reasons proving the human landing on the moon is a lie. Moreover, many believers in the spherical Earth theory acknowledge that landing on the moon was one of the greatest human lies.

One of the most famous mistakes is the photo of the spacecraft that carried the astronauts and landed on the moon. It is in the history section in NASA website. If you look at the photo, you will find that the spacecraft is made up of paper and knit by a lot of adhesive tapes. Can you imagine that such paperboard spacecraft endured all the pressure and the like! This is in addition to the flapping of the American flag. How could this happen while there is no air there? Also, what about the movement of astronauts and spacecraft? Yet, it may be said that the flag did not flap and that there was a fixing material on it when it was brought back, and so on. Believe me, you will find many such discussions on the internet.

Also, there was no dust on the machines' bases. The moon's surface is not smooth or paved to argue that the dust was not seen spattered there because of the surface's smoothness. Again, why do we not see stars in the moon landing photo? They stated various reasons, but one of the most important reasons was that the camera was focused on the moon's surface and the astronauts who landed there, not the stars. In addition, although the photos were of different spots of the moon, their background is the same! Are all the backgrounds so similar to such an extent, or were those who took the photos and played with people's minds lazy and thus used the same backgrounds?

Also, there are some photos where the shadow appears at different angles, so if the only source of illumination on the moon is the sun, why were there different angles of the shadow? Does this not indicate that there are several sources of illumination that emitted rays from several spots that led to such differences in the angles of the shadows?

And how could the astronauts manage to get through the Van Allen radiation belt? Their answer was that they of course chose the right time to pass through it. Some may also claim that this belt is "a flame of fire and smoke," and the other would say they penetrated it "with power and authority!"

Anyway, according to the Russian news and media reports, the Russian space mission will work hard to prove that the Americans landing on the moon is a giant lie! This news spread out at the end of 2018.

31

IMAGES OF THE EARTH

WHAT ABOUT ALL these images of the Earth, and not just the Earth but the distant planets as well? The answer is that we, until the present moment, do not have even one real image of the whole Earth, that is to say an image that shows the full shape of the Earth. All of the existing images are fabricated or as they are called "composite images." Therefore, all NASA images that broadcast live images of the Earth do not show the entire Earth but parts of it, with some curvature at its edges. You may say that this curvature is because the Earth is spherical, but I would argue this is not true. The curvature shown in the images is due to the use of a fisheye lens. You can use this lens and try it yourself, and you will find curvature in the photos you take of flat surfaces.

Many of the images of the Earth published by NASA contain disparate differences, a matter that reflects they have gone further in deluding people. However, this has been discovered experts and specialist in photography and computer alteration. Such images are wildly spread on the internet. Besides, it should be noted that although there are dozens of satellites from different countries, they all do not have even one live image of the whole Earth. Does not this mean there is something suspicious?

In fact, I find it strange that whenever they launch a space rocket launched, they photograph the rocket from its sides and behind it. I wonder, why do not they photograph what is in front of it and broadcast that live? And what about space debris or waste? They say there are more than 500,000 pieces of this waste that spin in the Earth's orbit with speeds of up to 17,500 miles per hour. Where are the images of such waste captured using modern techniques? What is that kind of survival skill required for satellites and astronauts to not collide such debris?

Also, the images of Google Earth and Google Maps are not real, but assembled ones. If you zoom out with the lens to see the whole Earth, you will see that they are not live images, but composite ones that have been assembled and altered by computers. You will notice that there is no day and night in the same image, but always the day only, to give the users a better experience. Also, whoever notices the coordinate shape of clouds, and the like, will surely notice they are composite images. However, when you zoom in with the lens to see the streets, you will find them seem real, which makes the Earth look spherical when zooming out on the image. This is due to their viewpoint that the Earth is spherical. If they held that the Earth is flat, they would make it look flat when zooming out with the lens and make the images seem real when zooming in. They could also make the Earth seem cylindrical or anything like that.

Dear reader, they claimed that they landed on the moon fifty years ago. Can they not use advanced technology to photograph thousands of images of the Earth as they claim? The real images should have been spread out. People nowadays love photography so much and take photos of everything, be it great or trivial. Are there none of those people in the staff of the space station and those working in this field who ever eager to photograph the whole earth. The absence of any real image proves that they have not yet discovered the whole Earth, nor have they realized its boundaries yet.

This aforementioned point is also the answer to why there is no image of the flat Earth. You, like any other person, may ask why there is no image of the flat Earth? The answer is that till the present moment, man has not explored or discovered the whole Earth, and the power of humans is actually too limited to do so, as they have not yet even reached the edges of the earth. The Earth is much greater than they imagine.

The Earth faces the sky, and just as they have not floated but in a small part of the sky, they also have not explored but a small part of the Earth.

Since humans can only fly in the atmosphere of the sky, as I said earlier, they will not be able to take non-fabricated and real live photos of the whole Earth. So, anyone who wants to photograph the whole Earth must go above and beyond the nearest heaven/ the sky, which is called in the Quran "a protected ceiling," but how can ever they be able to do?

32

PHOTO OF A BLACK HOLE

THROUGHOUT THE LAST two years, the initiatives, projects, films, and photos are growing in number in order to prove that the Earth is spherical or to verify modern science theories. A few years earlier, I used to tell my brothers that media men will focus on such issues as they begin to grow concerned with the audience. This is because the idea that the Earth is spherical began to spread widely and found acceptance amongst many people. There are thousands of countless clips and essays in this regard. Thus, siding with the idea of the spherical shape of the Earth, the media would do its best to refute the idea that the Earth is flat and claim that it is spherical. But as Allah said, "**So they will spend it; then it will be for them a [source of] regret; then they will be overcome**" [Al Anfal:36]. In this regard, in April 2019, an alleged photo of a black hole was circulated as the first ever photo of a black hole, claiming that such photos substantiate their theories.

However, let us study the case. Science tells us that the closest black hole to us is *Monocerotis v616*, which is approximately 3,500 light years away from the Earth and its size is 9–13 times that of our sun.

Well, let me ask you, "If you were in a car with your family and you were holding a camera directed toward the moon with the intention of taking its photo, then the car driver moved it, would the moon disappear from the camera focus or not?" Of course, it would. Then, you would have to refocus the camera on the moon, or else you would not be able to take the photo because of the moving car.

Just imagine the following! Scientists say that we are living on the Earth that moves with a speed of 1,000 miles per hour around its axis and 67,000 miles per hour around the sun. In addition, they say that the sun moves in the galaxy with an approximate speed of 500,000 miles per hour, and that the solar system moves with a speed amounting to nearly 515,000 miles per hour, given that the galaxy expands enormously. That being said, they claim that they took a photo of a black hole. They did that without having any harmful impact for refocusing their camera on the black hole. They had to do so to make the light surround the black hole to be able to photograph it. If they hadn't done so, the light wouldn't have escaped the black hole. If that had happened, darkness would have prevailed, and no photos could have been taken. In short, they worked a miracle.

Certainly, they couldn't have deceived people using a real photo. People are way too smart to be deceived by such obviously flawed tricks. The scientists rather said that it is a radio image. Though they agreed on that, I need not argue with them. It suffices to read what they themselves said about the way they took it to know that you cannot prove that they are true.

Remember that those same scientists have no photos of the sun system closest to the earth: Alpha Centauri, which is 4.5 light years away. In fact, we have no photos showing the alleged sand specks on the moon. By this, I mean a true photo, and not one designed by computer programs. Similarly, we do not have photos of the satellites that are so close to us if compared with the alleged black hole. Also, if you demand a photo of a space station, you will be presented with nothing but a small, blurred photo of it. Each and every year, scientists disagree on the day that marks the beginning of the Holy month of Ramadan. If they really have these great capabilities through which they could work the miracle of photographing a black hole, why can they not photograph these celestial bodies close to us, the act that would undermine the idea of the flatness of the Earth?

What about the Hubble space telescope? This telescope sends wireless frequencies that travel long distances to register the celestial bodies' size, revolving speed, and orbital relations with other celestial bodies. It also helps recognize their constituents of water, gases, other constituents, and the like.

Imagine the monumental dimensions they mention in the scientific periodicals: light years and billions of miles. Imagine that they can measure them precisely through radio frequencies that cross the cosmic dust, thousands of space wastes, space belts, asteroids, and many bodies floating in the space as they allege without causing these frequencies to lose their path or deviate away from the antenna waves, and they finally succeed in recording the physical phenomenon that takes place in space billions of light years away.

Given that the width of our constellation is over 46 billion light years, which equals about 276 miles with 21 zeros on its right side, can you imagine how colossal this number is? Thank Allah that scientists admit they aren't live photos. Scientists admit that they are radio photos and many people nonetheless think that they are real photos. What would have happened if scientists were carried away with their lies and said these were true live photos?

33

CONSPIRACY

N ow that you have reached this page, you may be wondering, "After all these pieces of evidence that prove the Earth is not spherical, why is the idea of the spherical shape of Earth still accepted and popular these days? Do you believe in the existence of a cosmic conspiracy?" I answer them by saying that Allah knows best, but I do not exclude the idea of a conspiracy or of propagating a lie. To say that there is no conspiracy is, in my opinion, far from correct. Satan threatens people that he would lead them astray and would spare no effort to deviate them from the Path of Allah Almighty. I do not think this act of misleading is confined to his instigating a person to be disobedient to his parents or to sow the seeds of disagreement between man and wife. Satan aspires to lead all humanity to hellfire, if possible, and what is the easiest way to do that? To lead them astray from the fact that Allah exists so as to deny this fact, or at least to disbelieve in His Names or underestimate Him.

Theories such as the Big Bang theory and the spherical Earth are premises that may lead the contemporary man to atheism, once he is deep to his neck in them. Alternately, they may have no effect on him at all. Don't you notice that the Big bang theory insinuates that the universe came to existence by accident or in a random way, and that there is no act of creation of a system that you find creative, wise, and abled? Don't you see that scientists' theory suggests that the Earth is no better than any other inferior planet that was similarly formulated, and that the sun and the moon are no better either, for there are millions of similar bodies? Then, the theory of evolution emerged to take its part in deviating people from the Divine story of creation mentioned in the Holy Books. According to this theory, the universe is the result of an evolutional process. The same applies to the theory of gravitation, which explains everything and cannot be tested. Given that, scientists no longer need a god, as one of the senior scientists actually said.

Compare this theory with the fact of Allah's creation of the heavens and Earth, making man vicegerent on Earth, giving man centrality, and that Allah subjugated day and night and moon to serve man. Allah subjugated the sea and land creatures and the birds in the sky to man. The mental and psychological impacts of each of these ideas is completely different. One of them is rife with systematic acts and significance, while the other is full of chaos and fabrications and lacks all the lofty meanings.

WHY DOES THIS LIE PERSIST?

If they hadn't landed on the moon, why did they propagate that they did? It is because they have no answers. Apparently, they wanted to land on the moon and when they failed to do so, they could not declare their debacle in public. This would have put them in many political problems. It would suffice if you just take a look on the budget of NASA's space programs or the space programs in India or China and other countries. It is a gigantic number.

Why do they spend all that money? Because it is an easy way to deceive people and steal their money. If they repeated day and night that the Earth is spherical, that the resources on it are limited, and that there are other planets similar to ours, the common man would look forward to exploring the unknown. He might even get frightened to the extent of donating money for this cause. He might encourage such programs in imitation of others.

Billions of dollars are spent on such programs even in countries with a high ratio of poverty like India. They allocated a budget exceeding a billion and half dollars for their space program. Can you imagine that hefty amount? Couldn't they use such an amount of money to combat poverty? A quick study of some of the states' budgets allocated to space programs may lead one to really doubt the validity of the spherical shape of earth. For instance, in USA, the allocated budget is more than 20 billion dollars and in China it amounts to 11 billion dollars. Europe spends around 6 billion dollars while Russia spends 3 billion dollars on such programs. Thus, the total state budgets allocated to space programs amount to about 450 billion dollars. How are these allocations spent? What else can be done by humanity with such money instead of studying the fictitious space and the constellations millions of light years away, or taking a photo of a black hole millions of light years away from us? Unfortunately, people do not know where their money is spent, yet they spend it imitating others and in fear and aspiration, and Allah is the One Whose Help can be sought.

Did the countries benefit from spending all these amounts of money in improving people's life or did they end up with making people live amidst endless billowy waves of lies and illusions? Don't the people living now deserve to know the truth, regardless how bitter it is instead of spending all this money on programs of space exploration? Did these programs help decrease or increase the number of wars?

States look forward to the far future and do not realize the risk of the problems suffered in the various human, security, and cosmic fields. Instead of spending money on the far future to save humanity then, it should be spent to save humanity now. If you save humanity today, humanity will be automatically saved in the far future. Claiming that we do not have enough time is but a means of sucking the money out of the pockets of both nations and states. This is one of the fundamental principles of strategies well known to those who study strategies, and I specialized in it due to my work in the field for so many years, praise is due to Allah.

The issue has so many political dimensions. I am not saying that scientists colluded to lying, but some countries participate in these space programs for political or economic purposes and the like. They do not participate to verify the theory of the spherical shape of the Earth. Rather, they participate to show their people that the state is a developed one or to attract investments and for other strategic purposes to be achieved by these programs. For instance, if USA has a space program managed by NASA and then India set up a space program, who would be the expert to be referred to, consulted, and resorted to in matters of training? For sure, it would be those who have expertise. Thus, NASA would benefit. Who would also manufacture the rockets, equipment, and satellites? The space sector is considered to be a vital sector these days, and the financial beneficiaries of this sector are numerous. Each of them has a special purpose. Thus, it is crucial for them to propagate the fact that the Earth is spherical. If the Earth is flat, they would lose many opportunities to reap profits. This is obvious, but it does not mean that they do not collude to lying. I neither prove nor negate that, and Allah knows best.

Do you know how much were the American government's expenditures on NASA since its establishment and up till now? Can you guess? They spent more than 600 billion dollars. If you converted this amount to Saudi Riyal or Emirati Dirham, you would grasp how much was spent on this agency till this day. Do the calculations yourself and then think of the reasons! Of course, you may wonder why they spent such a stupendous amount of money. This is simply because they could increase the chances and multiply the channels of revenues. They were the ones who said that the Ozone layer is in danger and, thus, they forbade the construction of any space stations without their own prior permission. If a given country had this permission, what would be the next step? Would it build the stations, manufacture the equipment, and gain experience alone? No, the state would consult the experts. Who are the experts? NASA and its affiliates. It is certain that the employees of these countries would need training, outfits, and equipment. Where would they get them from? They would buy them from agencies and partners, because the latter regulated space and developed laws for the space programs. Since the Earth is an issue that concerns all of us humans, they plead through the media to the countries and the people living on it by saying, "Our planet earth is in danger. Our resources are limited. We have to find an alternative. We have to save the earth. We have to protect our strata." When the states and peoples agree to provide assistance, they ask about what should be provided. Then, the expert states would say to them, "Give us your money and efforts!" You can imagine the magnitude of the amounts of money paid for such programs and the elephantine revenues of these agencies and space organizations, which are monopolized by some states that manipulate the people's money.

They say that with these amounts of money, they seek to save people in the future, to achieve man's happiness and protect him and the place where he lives from any harm. I can retort saying that if all these amounts of money were spent to face the real challenges like poverty, illiteracy, and the like, we would have minimized them largely, and we would have put an end to the challenges facing humankind. Yet, it can all be attributed to the humans' psychological defeatism, for they could have wondered, "What did the countries trying to keep pace with USA and other advanced and monopolizing countries achieve? Can these countries impose their hegemony on the whole world as USA, Russia, and others did? No! These countries monopolized these fields. Accordingly, no matter how great the advances the other countries achieve in this regard, they would remain subservient to these countries, for they receive various services from them. These countries put agreements, protocols, and memoranda of understanding to abide by the regulations they developed regarding the cosmic and environmental challenges.

We have to wonder about the secret behind the fact that official organizations like UN uses a map resembling that of the flat Earth commonly used amongst ordinary people instead of using the map of spherical Earth? Scientists would answer saying that they deploy the present map of the Earth characterized with Azimuthal equidistant projection that shows all the states without preferring any one over the other. I would not elaborate on the emergence of the flat Earth map on many logos; that is not my main interest. It might be mere coincidence. However, it is undoubtedly interesting to research for those interested in these matters.

As I mentioned before, we do not have a single real time photo of the whole earth up till this day. All the photos we see on TV and find on the internet are unreal. They are computer-designed photos and scientists are not ashamed to admit that.

Moving to NASA's live broadcast on the internet, I dare you to find a live broadcast of the whole Earth. You will always find a broadcast of a mere part of it that is a little deviated. This can be easily done using today's modern programs, such as through the use of the fisheye lens and the many other filters. Just search the sites of the alleged space programs and satellites and look for a live true photo of the Earth through their sites. You would not find a single one. I kept searching for a few days through the programs of USA, India, China, UK, and Russia. I tried to search in their official sites for a live broadcast of the whole Earth, or a real photo that is not drawn or imaginary of the whole Earth, only to find none. This substantiates the existence of a sort of conspiracy. There is no third option. It is either a conspiracy or a collusion amongst nations, and both are bitter options.

Yet, today, after the advances made in technology, I cannot exclude the possibility that they could have deceived people by broadcasting live-like photos of a spherical Earth using modern techniques like virtual reality. Those who visited some of the top governmental facilities or those dealing with modern technology and future surely have seen something of the like and can realize how computer-designed photos can look like live photos. Even in video games like PlayStation games and what came after, there are unreal clips that look amazingly real due to the precision of drawing and animation.

I pity the upcoming generation. They would not be able to easily distinguish between real and fake photos because of these modern techniques. This is a dilemma. Lies would spread widely. We cannot exclude the possibility that some of the Christians would use these theories to make photos of Christ in Heavens greeting them, sending his blessings upon them, or walking through heavens. All of this is possible today because of the modern photography techniques.

In fact, there are many clips on the internet today that prove the mistakes, lies, and falsifications in the alleged photos of the Earth. You can look for them and find them easily, thank Allah. However, I do not want to refer you to these photos, clips, and experiments. All these are stuff that can be fabricated.

If you wondered why we don't have a photo of the flat Earth showing its ends, the answer would be easy. We did not reach the ends of the earth yet. We didn't even explore it or set ourselves free from its hold. We did not go out of the Earth boundaries yet. The flat Earth is much bigger than we imagine. Even I do not know what is beyond the inhabited Earth. Are there other lands, or it is only the wide sea? Only Allah knows. This is because man's abilities at this moment are incapable of passing beyond the regions of skies and land, and man did not penetrate the Earth yet or go out of the skies above us. That is why we have no real photo of the whole flat Earth, and man cannot take a photo thereof. This is because the Earth spreads infinitely with no end known but to Allah Almighty.

Amongst the bad things used by some of the means of the media through their different documentaries or programs is that they picture the advocates of the flat Earth theory as psychos or people suffering loneliness. They are presented as persons who do not feel loyal to society as a whole or care about it. They present them as persons who have an unstoppable desire to be in the spotlight because this gives them safety. To prove this, media men may meet a psychiatrist, an expert on the human behavior or the like to discuss psychological complexes. Unfortunately, this is the method of the weak who—on failing to produce evidence—resort to any method that destroys the opposing idea. This is because the weak party cannot coexist with a different point of view. In fact, those weak persons are the ones who need psychological treatment and not those who advocate the flatness of the Earth. Deploying mental or psychological intimidation against them or stigmatizing them with madness or illness or the like would not make any advances in real scientific research. Instead, media men should have compared the evidence of flat Earth and that of the spherical earth and compared them fair and square without siding with either view. This way, humanity would progress and move ahead. Excluding all the opposing views and attempting to hide them by any way was not one of the characteristics of those pursuing the truth. Those who advocate the truth are fearless, even if falsity prevails, because they know truth is like light. It lightens darkness, no matter how wide it prevails.

I do not deny that amongst those who advocate the flatness of Earth are some having psychological disorders. Yes, this can happen, as psychologically disordered people can be found anywhere and the existence of some of them within a given category does not make this whole category or all those advocates of the idea psychologically ill. This is a grave mistake deployed by media men, especially those who endeavor to broadcast anti-Islam programs. This is obvious and known to the Westerns, Arabs, Muslims, and non-Muslims as well. Distorting the image is an act of those who lack consciousness of themselves or others. Thank Allah, I, as a Muslim, have the Holy Quran that refuses these twisted methods to refute what is true. Contemplate this entertaining Quranic verse that urges the Muslims to be patient. Allah, Almighty said, **"Rather, We dash the truth upon falsehood, and it destroys it, and thereupon it departs. And for you is destruction from that which you describe"** [Al Anbiyaa:18].

What I find irritating is that whenever they come up with a theory, I find that the apparent meaning of the Holy Quran runs counter to it. This is the same technique by which weak Hadiths are discovered: by comparing them to the apparent meaning of Quran. This way, the flaws in the Hadiths and their weaknesses are unveiled. Allah, Almighty said, **"And at the earth—how it is spread out?"** [Al Ghashiya:20]. But scientists say it is spherical. Allah Almighty said, **"And Allah has made for you the earth an expanse"** [Nuh:19]; and He said also, **"Is He [not best] who made the earth a stable ground"** [Al Naml:61]; but scientists said that the Earth revolves. Allah tells us, **"[He] who made for you the earth a bed [spread out] and the sky a ceiling"** [Al Baqarah:22]; and Allah said, **"And constructed above you seven strong [heavens]"** [Al Naba':12]; but scientists hold that there is nothing but vacuum and emptiness. Allah said, **"And He restrains the sky from falling upon the earth, unless by His permission"** [Al Hajj:65], but they say that the Earth is in space and not beneath the sky. Allah also said **"and a garden as wide as the heavens and earth, prepared for the righteous"** [āl 'im'rān:133]. They said it is a mere atom floating in the space! Allah told us that heaven has no rifts, and thus said, **"Have they not looked at the heaven above them—how We structured it and adorned it and [how] it has no rifts?"** [Qaf:6]; and also has no breaks and thus said **"[And] who created seven heavens in layers. You do not see in the creation of the Most Merciful any inconsistency. So return [your] vision [to the sky]; do you see any breaks?"** [Al-mulk:3]. They told us about black holes. Allah told us that **"it is He who sends down rain from the sky"** [Al-an'ām:99]. They said steam from the Earth. Allah said that you will not pass beyond the regions of the heavens and the Earth except by his authority, and thus said, **"O company of jinn and mankind, if you are able to pass beyond the regions of the heavens and the earth, then pass. You will not** pass except by authority [from Allah]" [Ar-Rahman:33]. They said they managed to pass with the power and authority of science. Allah said, **"There will be sent upon you a flame of fire and smoke, and you will not defend yourselves"** [Ar-rahman:35]. They said meteoroids float in the space. Allah said, **"And the sun runs [on course] toward its stopping point"** [Yasin:38], and the Earth is fixed. They said the sun is fixed and the Earth spins. Allah said, **"Blessed is He who has placed in the sky great stars and placed therein a [burning] lamp and luminous moon"** [Al-Furqan:61]. They said, the moon is not luminous, but it reflects the light. Allah said: **"that which I created with My hands"** [Sad:75], and they talked about evolution!

I say this is an explicit and blatant contention to the words of Allah, and they said, "Who are you even to comprehend the words of Allah. It is not your domain, so just keep off!" I do not think I will be blamed if I feel intellectual loneliness.

It is really a fact that most people believe in what is contrary to the apparent meanings of the Quran. Praise is to Allah, Who said, **"And if you obey most of those upon the earth, they will mislead you from the way of Allah. They follow not except assumption, and they are not but falsifying"** [Al-an'ām:116], as this verse truly relieves and comforts the hearts of the poor like me who do feel intellectual solitude.

POSTSCRIPT

H AVING BEEN PATIENT with me, all I wish is that this book would cast doubts in your mind regarding what the media try to install in our minds, day and night. I just want this book to urge you to search for the truth yourself without lending your mind to others or borrow their minds to think in your stead. Seek the truth, pursue it, dig for it, examine the views and proofs, compare between them, select the reasonable ones, give precedence to the powerful ones, exclude and rule out the weak ones, and form your own view. You have to do a lot of things to reach a result that will satisfy you.

I am sure that if you were patient enough and read the book until you reached this page, I have achieved my goal of making you eager to search and learn more about the issue of the shape of the Earth, whether it is flat or round. Since the question of the shape of the Earth and whether it is spherical or flat is a subject of controversy and conflict between people nowadays, what shall be done to reach the truth in this regard? Allah, Almighty, has guided us to the solution through the following verse, **"O you who have believed, obey Allah and obey the Messenger and those in authority among you. And if you disagree over anything, refer it to Allah and the Messenger, if you should believe in Allah and the Last Day. That is the best [way] and best in result"** [An-Nisaa:59].

Dear reader, since you have learned that people have disputed over the shape of the Earth, refer to the Quranic verses and the hadiths attributed to the Prophet (peace and blessings of Allah be upon him). Take a moment and read the Quran meditatively from the beginning to the end, without letting anyone to influence you, and judge for yourself whether the Earth is spherical or flat. I will leave it up to you.

I advise you, dear reader, to pray to Allah when you look into the issue, along with having a sincere intention to know the truth and spread it out. I do not remember even one night when I have not prayed to Allah since I started writing this book with the following, "O Allah, show me what is true and help me to follow it, and show me what is false and help me to avoid it." If I have been granted success to arrive at the truth, the fact, and what is right and beneficial, it is by Allah's Grace. If I have mistaken or erred, it is due to my fault, my sins, and by the Shaytan (Devil). Allah is the only one Who guides to the right path, and to Him is the return of everything. Peace be upon you!

APPENDIX 1

PREFERENCE FOR THE TRUTH

OVER THE CREATION

P EACE AND BLESSINGS of Allah be upon you. This is my first entry on the blog, this year. I would like to write my first article to express my thoughts concerning some important issues that may seem silly for some people.

I chose the title *Preference for the Truth over the Creation*, since I have been affected by the title of the book of Ibn Al-Wazir Muhammad ibn Al-Murtada Al-Yamani. There are other reasons for choosing this title for my article as well, but there is no need to mention them here.

I would like to write about the shape of the Earth, an issue that perplexed many people more than 1,000 years ago. Is the Earth spherical or flat? The question may seem strange, and even questioning that may be considered idiocy and ineptitude. However, the question is not so, according to many scholars and skilled scientists. In fact, I have discussed such questions with some fellows on the internet for more than fifteen years. The question has been raised every now and then, and it has still not been resolved, at least for me. I prefer the thought that the Earth is flat and that the spherical Earth is nothing more than a theory.

You may wonder what urged me to research on this question. For me, there were two basic matters: First, the Holy Quran, and second, what is now called the Bible. As for why I write about this issue now after I had quit it for many years is because some Arab writers wrote silly articles that provoke those who hold that the Earth is flat. Those writers, who should have been educated and respected, followed bad practices that they used to prohibit in their writings and articles.

They, for example, accused those who say the Earth is flat of being imitative and backward. (Of course they mean that they imitate Ibn Baz or Ibn ʿUthaymin, but they do not understand that the Fatwas—religious rulings—of Ibn Baz and Ibn ʿUthaymin were about the movement of the Earth, sun, and moon, and not about the question of the shape of Earth and whether it is flat or spherical.) However, I do not follow one certain intellectual school or thought. I am not a Sufi, Salafi, Muʿtazili (a follower of the theological school of Muʿtazilah), or Ashʿari (a follower of the theological school of Ashʿaris). Thanks to Allah, I am not inclined to support a view other than the other, and therefore, I am not afflicted by "label phobia,"

Many intellectuals who mock those who hold that the Earth is flat have no knowledge at all about this question. They mainly imitate the scientific community and the dictations of mass media as if they were revelations sent down by NASA and the other space agencies. Moreover, such writers accuse others of practicing intellectual terrorism and extremism while they do the same. They terrify those who maintain that the Earth is flat by saying all the world is against them or that they are only a small band, or that the such-and-such government or the Great States plan to build a society over Mars, invade the space, and send a space ship to such-and-such planet.

The poor curious person who fears to undergo this state stops searching and follows such writers under the fear of being an outcast in the society or being mocked on account of the intellectual terrorism practiced by the advocates of tolerance, coexistence, and acceptance of others.

For this reason, I decided to write a little to make people curious to look for the truth, and, thus, the serious person will look for the answer to this question, freeing himself of the radical restrictions of imitation that may hinder him.

Let us look at the core of the question, and I will try to give you some proofs indicative that the Earth is flat. However, I would like—before that—to point out that the proof should be presented by those who hold the theory of the spherical Earth, because they are proponents of this new theory that contradicts the customary belief of most people of the world and contradicts common sense. However, I will tell you what I have.

As for the movement of the sun and moon and the fixedness of the Earth, this is proved in the Quran and is crystal clear for every sensible person unaffected by the other perspectives. This is the opinion of Sheikh Ibn 'Uthaymin, may Allah be merciful to him, and he was correct. Let people say what they want to say; truth is more worthy to be followed. Didn't Allah, Exalted and Glorified is He, say, **"And the sun runs [on course] toward its stopping point"** [Yasin:38]? Note that He did not say "you see the sun runs" or the like, but He rather stated a true fact. He said, **"And the sun runs [on course] toward its stopping point."** He also said, **"And [had you been present], you would see the sun when it rose, inclining away from their cave on the right, and when it set, passing away from them on the left"** [Al-Kahf:17]. He said so because this question is a perspective for man. He, Glorified is He, said, **"Indeed, Allah brings up the sun from the east"** [Al-Baqarah:258]. If you considered the opinions stated about the difference between *Ja'a* (came) and *Ata* (came) in meaning by Dr. Muhammad Shahrur or others interested in the meaning of the Quranic term, you would find them completely support the idea that the sun moves, and the Earth is fixed. Allah, Exalted is He, said, **"Is He [not best] who made the earth a stable ground"** [Al-Naml:61]. This can be interpreted for any party if we put into consideration the rules of interpretation they developed for themselves.

For me, most of the verses about the Earth refer to the fact that the Earth is fixed (in its natural state without earthquakes), while the sun and moon are moving. As for the origin and reality of the Earth and whether it has movement other than the scientific movement they claimed, this is possible. Why not? For example, if the whole mass of the Earth had been above the water, it might have been moving above it like ships, with the winds moving it to the right and left, and without mountains that fix it. That's why Allah said, **"And He has cast into the earth firmly set mountains, lest it shift with you"** [Al-Nahl:15]. Understanding this needs wide imagination beyond the horizon of the imitators of the modern scientific community. However, I may be wrong, and Allah knows the right.

The third matter is that the sky is a canopy, and it is known that every canopy could be destroyed, in accordance with Allah's Saying, **"And He restrains the sky from falling upon the earth, unless by His permission"** [Al-Hajj:65]. This is totally in accordance with the idea of the flat Earth but is problematic with the perspective of the spherical Earth, as where is the Heaven in the globe? And how could the sky fall while we are a point floating in the space? Does the enclosing of the space refer to the falling of the sky? Does this mean the sky will fall over the heads of those who are in the southern hemisphere of the globe? What does this mean for whoever reads in the Quran who lives in the southern part of Earth that floats in the space? The direction of the sky will then be downward for those who live in the north. In the same manner, the sky will then be in the northern part downward for the people of the south. All these verses will be problematic for them and thus there should be interpretation. The well-known rule is that the apparent meaning of the Quran should not be changed except for a considerable proof. Where is here the considerable proof?

Such people do not consider the Hadith attributed to the Prophet, peace be upon him, proofs because they were narrated thousands years ago, and humans are subject to error, while they say that the images produced by NASA and other space agencies are not for lies and deception of people, even though they were proved to be liars more than one time. If the rules of science of Hadith and Jarh and Ta'dil (invalidating and validating narrators) were to be applied to NASA, it would have been regarded not trustworthy.

Besides, the proof of images is not to be taken into consideration because the photos may undergo serious manipulation and fabrication. Therefore, how could you interpret the apparent meaning of the Quran with proofs not established in terms of principle, science, and reason, while the fundamental and rational proofs are in support of those holding that the Earth is flat?

Where are the rising place and the setting place of the sun in the story of Dhul-Qarnayn in the Quranic Chapter of Al-Kahf? What does **"he found it rising"** mean? If it is just a perspective of the spectator, what distinguishes Dhul-Qarnayn from other spectators? I understood from the context of the verse that Dhul-Qarnayn reached what could not be reached by ordinary people. Does

Allah, Exalted and Glorified is He, in His Saying, **"Lord of the two sunrises and Lord of the two sunsets"** [Al-Rahman:17], and His Saying **"Lord of the sunrises"** [Al-Saffat:5], mean "the lord of the different perspectives?" Or, does He mean true and not theoretical sunrises and sunsets? Glory be to Allah! We need a great deal of discipline. I ask Allah to forgive us all.

How could the true direction of Qibla (Ka'bah-direction faced in Prayer) be determined if the Earth is spherical? Suppose the Ka'bah is in the middle of the top of the Earth while I stand in the middle of the bottom of the Earth, if I turn eastward, I would be turning toward the Qibla and if I turn westward, I would still be turning toward the Qibla. The same is true if I turn northward or southward. However, this does not occur in the real life of people, as every Muslim turns toward the Ka'bah, regardless of his place. Since the Earth is flat, we will have only one direction, and this is what happens in reality. Contemplate this evidence for it is a strong proof by which Allah may enlighten you.

Did Allah, Exalted is He, not say, **"Have they not seen that We set upon the land, reducing it from its borders"** [Al-Ra'd:41]? Where are the borders of the Earth? The context of the verse indicates a certain meaning, and it may be a warning for the disbelievers that their land will be taken from them. I know this well, but I want to say that Allah proved the Earth has borders. What comes to the mind are the ordinary, known borders, but after interpreting the verse allegorically, we find that it refers to the lands of the disbelievers. It is well known that the Quranic verse should be interpreted as meaning what it is totally referring to, if it is existent, and this applies to the flat Earth, because the flat Earth, unlike the spherical Earth, has borders. However, if you apply it to the flat Earth, you will find it has more than one meaning, but if you apply it to the spherical Earth, you will find it has only one meaning that may raise some questions as well.

As for those citing the prophetic Hadiths as evidence, we ask them what they say about the Hadith stating that the sun prostrates under the 'Arsh (Allah's Throne), or the Hadith stating that the Lord descends to the lower Heaven, and other Hadiths that seem problematic for those who claim the Earth is spherical. I have my special reservations, as they need more detail in terms of the Sanad (chain of narrators) and the Matn (text of the Hadith), but they are more in accordance with the idea of flat Earth.

There are so many proofs in the Bible that the Earth is flat; I counted about fifty proofs, while others claimed that they have more than four hundred sections indicating that the Earth is flat and that it has four corners.

The problem is that the Christians mock Muslims under the pretext that their book contradicts the modern science, as it states that the Earth is flat. Meanwhile, Muslims like Zakir Naik and others say that the Bible is a book that contradicts the modern science as it claims that the Earth is flat, contrary to the modern science.

The two factions do not know that the Bible and the Quran agree that the Earth is flat. Had this information in the Bible been wrong, Allah would have corrected it in the Holy Quran, as He did in respect of some misconceptions in the Bible that people misunderstood, such as the Lord's rest on Saturday— Exalted is Allah above feeling tired and weary. Most religions agree that the Earth is flat, so if you said that the Bible is not approved because it is distorted, I will say, "Bring me first the proofs of its being distorted and then we will discuss the question of the distortion of the Bible." However, why would people distort such information two thousand years ago and then the Quran would not prove the wrongness of this information? I will not bring the proofs from the Bible that the Earth is flat, for they are so many. You only have to Google them, and you will find them.

The Bible states that the flood drowned the whole world, and this was vilified by Dr. Adnan Ibrahim, may Allah bless him, in one of his speeches, and by many other thinkers, though the Quran does not state explicitly that Allah drowned the world with the flood. The information we get from the Quran is that water overflowed the mountains, and this information is enough to indicate that the Earth is flat, because if it was spherical, how could water overflow the mountains? However, the meaning is clear and understood if the Earth is flat. Do not tell me about the gravity and talk nonsense like children in blind imitation of what was said. You should rather look for the coherence of this theory and its suitability to interpret this topic and then you judge by yourself.

As I talked about the great scholar Dr. Adnan Ibrahim, I would like to point out that he is my fellow, and a unique scholar and philosopher. However, I think he has likely been affected by the scientific community and disregarded that they may be wrong, so he did not carry out research as he does in other issues. For this reason, he is—as well as the great scholar Bassam Jarrar—with the vilifiers against those holding that the Earth is flat. I am not more knowledgeable than any one of them, for both of them have a great deal of knowledge and action, all praise be to Allah, but I think it is not difficult for me to disagree with them and the scientific community in this topic of the flat Earth. However, I ask Allah to grant me and them success and guide us to what pleases Him. Let us go back to our topic.

The people of Noah, peace be upon him, knew about the seven Heavens, which are difficult to understand in case the Earth is spherical. However, if we considered the Earth is flat, it would be easy to understand the seven Heavens one above another. Besides, Noah argued with his people saying: **"Do you not consider how Allah has created seven heavens in layers"** [Nuh:15]. If they knew nothing about them, why would they be asked about them? However, what are the seven Heavens? Allah knows better, I do not know for sure. However, the verse may probably refer to the well-known seven planets, each moving in their respective Heavens. They may be a sign of the layers of Heaven. They are not in fact one layer; they are rather layers, one above another, though anyone who looks at the Heaven thinks they are one layer. Imagine the rainbow. Try to draw it on a paper and color it with its different colors, and then put in each color a small star. What will you see? You will see each star in a special orbit or layer. I think, and Allah knows better, the same is true with the seven Heavens mentioned in the Quran.

Besides, Allah, Exalted and Glorified is He, says in the Quranic chapter of Noah, **"Then We opened the gates of the heaven with rain pouring down"** [Al-Qamar:11]. Where are the door of Heavens today? But if we said that the Heaven is a canopy or—as the Bible calls it—Firmament, the matter can be clarified as follows: Allah opened the doors of Heaven so the heavenly water which was over the sky fell, and water gushed up out of the earth, overflowed, and reached the mountains. Yes, my fellows, I opine that there is water in the Heaven and under the Earth and surrounding it. Yes, you may believe I am insane, but I am, praise be to Allah, content with this view. It may take long to explain this as this is not all I believe in.

As for the refutations to the arguments supporting the theory of the spherical Earth, they are so easy, but I do not seek now to refute their arguments one after the other for this needs a whole book, and I do not want to take more than half an hour to write this article. Among these silly arguments promoted by the proponents of the theory of the spherical Earth is that it is an idea that dates back to the time of the Greek philosophers and that the early Muslims adopted the same view. However, those proponents do not know the meaning of "some," because not all philosophers but most of the pre-Aristotelian philosophers maintained that the Earth is flat. Also, some post-Aristotelian philosophers as well as Muslims maintained the same. Most Muslim scholars hold that the Earth is flat, with the exception of a few of them. Most religions hold that the Earth is flat, and the idea of the flat Earth was promulgated in China only in the sixteenth century. However, being supported by the majority cannot be used as an argument by any party; it is rather the argument of those who lack argument. As I mentioned before, to state the proofs of the flat Earth and refute the theory of the spherical Earth, I need a whole book. I do not think I will do so until I finish more personally important issues.

The sun and the moon are much smaller than their size reported in the scientific communities. The moon does not reflect the rays of the sun but is self-luminescent. The night and day existed before the sun and moon. Does it mean that light was existent before the existence of the sun? Yes, I believe so. But does my statement negate that the sun illuminates or is luminescent< No, it is luminescent. It is the sun that rotates and moves while the Earth is fixed in its sensible natural states. The 'Arsh may be as it is commonly thought, but the idea that it may be a set of the Heavens and the Earth, that is, the cosmic egg, as well as the heavenly dome and the inverted earthly dome and the water under it, is the bomb! That is why there is no mention of the creation of the 'Arsh in the Quran, because Allah mentioned the creation of its details.

The Earth is not spherical, but the seven Heavens and the Earth are a set, and if you looked at them from outside, the universe may seem spherical or like a bubble floating over the water. The movement of the mountains that float away like clouds seems to me to be in the water, and the mountains are like the anchors of ships. If you consider the Earth to be a ship or a boat, and then you set the mountain on the Earth such that the mountain would break through the ship and reached the water under it, the mountain would then move like the movement of the clouds. The explanation of this needs several articles, but the intelligent will understand what I mean. The Earth is like the base of a tent, and the Heaven is the tent itself. Could there be a whale under the Earth named Nun? Allah knows better, and that is not difficult for Allah. Is that whale alive or dead? I do not know, and Allah knows better. Does the Earth have pillars and bases? I think so. If you asked me about my theory concerning creation, I will sum up some matters that I think are likely more correct than what the scientific community agreed on. Allah is the first and the last.

The first created object may be water, and this water might have the shape of a bubble, egg, or it might be surrounded by a bubble or an egg. Then, Allah separated the Heaven from the Earth after they had been one united piece and made between the water of the heaven and the water of the Earth a barrier, as He says, **"Between them is a barrier [so] neither of them transgresses"** [Al-Rahman:20]. Then, Allah brought the Earth out of the lower water, and it was surrounded with water, and after the Earth was brought out there came into being the smoke in the vacuum between the upper water and the lower water, and He made the seven heavens. Regardless of your perception of the 'Arsh, His 'Arsh was on the water, whether in terms of the traditional sense or the sense explained below. The water of oceans is salty because it was mixed by two Earths and the water of the Heaven is fresh because it had not been mixed with the Earth. As the Earth was on water, it was unstable and would sway, so Allah cast into the Earth firmly set mountains, and for this reason, mountains pass like clouds. The seven Heavens are layers one after another, and the seven planets may be signs of their respective layers. The paradise, I think, is above the Heavenly dome, or in the highest layer, and Hell may be under the Earth. The Earth is then spread out, that is, its extremities are high. The Earth has four high extremities and four angles, and there may be a king—source of wind—at each extremity. Some of the rivers and the Earth we live in are lower in level than the extremities. The Sun and moon are in the Heavens, and its level may be closer than the high extremities of the Earth. It is probably called the 'Arsh for all the creation I described. It has bases or pillars carried by angels. Around this cosmic egg, cosmic tent, or cosmic bubble, there is creation known only by Allah, Exalted and Glorified is He. All knowledge belongs to Allah. Is there air under the cosmic water? I do not know. However, all that I said may give a clue to understand Allah's Saying, **"and made from water every living thing"** [Al-Anbiya':30].

You may wonder why I use "maybe" so much. This is in fact a kind of politeness for it is only Allah who knows the whole truth. I do not admire the scientific community or others who think they have reached absolute facts. You may say, "How did you imagine this?" I will answer you, "I do not know exactly. It may be out of delusions, imaginations, Sufi speeches, and dreams. It may be a divine inspiration or a devilish insinuation." Only Allah knows the truth, and I ask Him to let me see the truth as (it is really) true and help me follow it, and to let me see falsehood as (it is really) false and help me refrain from it.

I can say that I can prove everything I mentioned with Naqli (traditional) proofs and the through hidden relations between various texts, as it is not beyond reason. However, it is simpler than the beliefs of most people today in the statements of the scientific community. The problem is that people are not accustomed to such writings and they think they are more complex than what they believe in.

I may come back to this article and update it at a later time, if I have enough time and enthusiasm. Otherwise, I will leave it unedited. I may write other articles, but I am busy with other matters, and I do not know what is destined for me.

All I wish is that my writings arouse your curiosity so that you may search by yourself, and you will realize that the question of the shape of Earth is more complex than to be a point of ridicule. However, do not accept this article as an academic one; it represents just some thoughts and reflections I wrote in a hurry. It is not meant for scientific discussion. Peace and blessings of Allah be with you.

APPENDIX 2

REFUTATION OF THE SERMON OF SHEIKH ADNAN IBRAHIM: AGENTS OF NASA AND THE SPHERICALNESS OF THE EARTH

Peace be upon you! Over a week ago, I published my article *Preference for Truth over the Issue of Creation*, which I wrote about the flat Earth to the account of Dr. Adnan Ibrahim. I found him address that question, hallelujah. It seemed that he did so because of some of the posts published on the internet about the flat Earth.

I felt sad and disappointed when I listened to the sermon of the professor, but I was still somehow happy and glad. I was sad because a great professor like Dr. Adnan Ibrahim addressed in his current sermon what he used to forbade in his other sermons. He used the same emotional approach that his criticizers used against him.

The professor, of course, knows the famous poetic line that reads:

> *Do not forbid a conduct you do*
> *What a shame if you do!*

Undoubtedly, the professor must have forgotten that while giving his sermons; and who among us does not forget? I was disappointed that Dr. Adnan did not probe deep into the issue of the shape of the Earth, as this is clear from the information he presented in his sermon.

What made me happy is his sermon. After I had looked into the issue, I said to myself that there is no barrier to the results I and many others had concluded except Dr. Adnan's genius and knowledge. It is true that Dr. Adnan Ibrahim is not the most knowledgeable in terms of cosmic sciences, nor probably in terms of Sharia sciences, but he is an encyclopedic scholar who has acquired many branches of knowledge. He is a great scholar whose opinion is to be reckoned with. That is why he is called the Reviver by the youth. I do not know a famous Muslim scholar among the youth known as much for his truth, knowledge, and power of persuasion as Dr. Adnan. I think he has such a great deal of these three qualities. For that reason, I waited impatiently for a sermon where he could speak about the shape of the Earth.

I used to say to myself, "If Dr. Adnan was to speak about that issue, he would surely talk about a new matter, unknown for me, or would introduce strong irrefutable proofs and facts." However, he just offered a very simple and typical reply that any person with a little amount of cosmic and Sharia knowledge who argues for the flatness of the Earth could refute. In short, the strongest reply I have been waiting for years proved to be very weak and unsatisfactory. After that sermon, I became sure about what I had chosen for myself. May Allah reward him well forever! I ask Allah to enlighten and guide him if he was to read these lines any day.

Before starting to refute his statements, I would like to point out that I will not refute the scientific proofs and experiments he had introduced, because they are all weak and tenuous and have been refuted in many posts and articles and even in books, so I do not want to waste my time with them. However, I am concerned with the traditional or textual (*Naqli*) proofs, because refuting each one of them needs more or less a whole article like this one.

Let us begin. First, Dr. Adnan, may Allah protect him, chose the same approach used by his opponents, namely, intellectual terrorism.

"The majority of the people of the Earth hold that the Earth is spherical, except a few people who assume that the Earth is flat."

"Muslim scholars have unanimously agreed..."

O man! How could you contradict all Muslim scholars?

"Imam Ibn Hazm said . . . "

"Imam Ibn Taymiyyah said . . . "

He magnified each one of those scholars he mentioned, as they do deserve. However, he forgot how he contradicted Muslim scholars in questions held by no one before him (though I tend to accept many of his juristic choices), and how he who assumed unanimity is a liar. However, the difference is that there were no proponents in the questions in which Dr. Adnan contradicted Muslim scholars, as most of the early scholars held different opinions than his. With regard to the question that the Earth is flat, it is totally different as many people and Muslims held the same opinion.

Second, the professor's zeal while assuming the unanimity of Muslim scholars is mixed with false information. Suffice it to say that Dr. Adnan assumes unanimity reporting the unanimous agreement of scholars in this question, although he did not probe well into this question. Who said that there is, essentially, unanimity among scholars that the Earth is spherical? I do not mean the mere allegations of unanimity, for these are so many. Suffice it to look into an encyclopedia for unanimities to realize the false allegations made by some scholars to support their doctrines. How many times did Ibn Taymiyyah and Ibn Hazm, may Allah be merciful to them, make such wrong allegations? I think Dr. Adnan knows that better than I do.

I mean where is the reality of unanimity in the question? We find that Al-Qurtubi, may Allah be merciful to him, while interpreting this verse: **"And the earth—We have spread it and cast therein firmly set mountains and caused to grow therein [something] of every well-balanced thing"** [Al-Hijr:19], quotes Ibn Abbas's saying about its interpretation "spread it over the surface of water." He also quoted Allah's saying, **"And after that He spread the earth,"** and, **"And the earth We have spread out, and excellent is the preparer"** [Al-Dhariyat:48], while refuting the allegations that the Earth is a spheroid. As mentioned earlier, this is the opinion of Al-Qurtubi. Thus, there were opinions and refutations concerning the idea that the Earth is spherical. We can find in the poem of Imam Al-Qahtani Al-Andalusi (d. 379 AH), these lines that concern us:

The engineer and the astrologer lied
They claim that the earth is round
They held the same opinion
Though the Earth is flat for the people of reason
Based on the clear saying of the Quran
That Allah spread it for the people and built the Heaven well

Don't you know that Ibn Kathir, and many other scholars, held that the Heaven was created before the Earth, and the Earth was spread out after that? How does this correspond with the modern scientific theories regarding the creation of the Earth? We can also find in the Commentary of Al-Baghwi that the meaning of the same verse, **"And the earth—We have spread it"** [Al-Hijr:19], is, "We have spread it over the surface of water." It may be said that what he meant is oceans. It may also be said that the whole Earth is on water, as I mentioned in my article, *Preference for Truth over the Issue of Creation*. However, what is the meaning of the statement of exegetes that "He spread it in length and width." How could this apply to the spherical Earth, particularly since what is meant here is the whole Earth and not a piece of land or a certain land, as is clear from the context.

I agree with Dr. Adnan Ibrahim that the meaning of Earth differs according to the context. That is true, but when Allah speaks about the creation of Earth versus the Heavens, what is meant is the whole Earth and not a certain piece of land.

Imagine, for example, when Allah, Exalted is He, says, **"and a garden as wide as the heavens and earth"** [Al 'Imran:133]. Does the term "earth" here refer to only Mecca, Medina or any other piece of land? Or does it refer to the whole Earth? The answer is quite clear. There are so many verses in the Quran with the same meaning as the aforementioned verse. However, if it were meant that the Earth is spread and extended only for the spectator, the spectator can say that he sees mountains and highlands and that the surfaces of the Earth are different, so how is it spread and extended? The only reply would be that it goes beyond the perspective of the spectator. Thus, it would not be a matter of perspective and would turn to the whole mass of Earth. If so, we would ask, "why did you perceive a certain mass and not the whole Earth?" Had you interpreted the verse according to its explicit meaning, you would have reached the true result?

In the Commentary of Ibn Atiyyah (d. 546 A.H.) on Allah's Saying, **"And at the earth—how it is spread out?"** [Al-Ghashiyah:20], Ibn Atiyyah said, "The apparent meaning of this verse is that the Earth is flat and not spherical." This is the opinion of scholars. Arguing that it is spherical is a statement not proved by the scholars of Sharia, even though it has nothing to do with the pillars of Sharia. It is well known who Ibn Atiyyah is, and he said that this statement is not proved by the scholars of Sharia. Would it have been unknown to him if the whole Muslim scholars had unanimously agreed that the Earth is spherical, and would he have made such an allegation? Of course, he would not ever do that. He who reads the commentaries objectively and reads the concomitant Hadiths in the same respect would find that the allegation to the effect that all exegetes—or hundreds of them—agree that the Earth is spherical is not true. Similarly, in the commentary of Al-Jalalayn concerning Allah's Saying: **"And at the earth—how it is spread out?"** [Al-Ghashiyah:20], we quote, "His Saying 'spread out' indicates explicitly that the Earth is a flat surface as is hold by the scholars of Shariah, and not spherical as claimed by astronomers, though this does not affect any of the pillars of Shariah." Jalal Al-Din Al-Mahalli is well known. The same view was adopted by Al-Tha'labi in his commentary.

Even when Al-Razi wanted to prove that the Earth is spherical, he said in his interpretation, "Some people used this as evidence that the Earth is not a sphere." Then, he refuted their claims and proved at least that there is disagreement concerning this issue, unlike what Dr. Adnan led us to believe by stating there was no disagreement.

Similarly, Imam Abu Mansur Abdul-Qahir Al-Bughdadi (d. 429 AH), author of *Usul Al-Din*, is one of the Muslim Scholars who hold that the Earth is flat. Explaining Allah's Name of Al-Basit, Imam Abu Mansur said, "Al-Basit indicates that Allah expands the means of livelihood for whoever He wishes and that He expanded the Earth. That's why Allah called the Earth 'an expanse,' contrary to the claim of some philosophers and astrologers that the Earth is spherical and not spread out."

Also, among the supporters of the theory that the Earth is flat is the well-known geographer Siraj Al-Din ibn Al-Wardi (d. 852 AH). In his book entitled *Kharidat Al-'Aja'ib wa Faridat Al-Ghara'ib*, he says, "The greatest mountain in the world is Qaf. It surrounds the Earth as the whites of the eye surround its pupil. As for what is beyond the Qaf Mountain, it is a matter of the Hereafter and not the worldly life." Qaf Mountain is reported in some narrations that we are not going to ascertain. However, Ibn Al-Wardi regarded the Mountain as a barricade around all the Earth. It resembles some of the new orientations in favor of the flatness of the Earth that negates the sphericalness of the Earth that floats in the space like an atom. However, I would like to note that the aforementioned book is sometimes attributed to his grandfather, Judge 'Umar ibn Al-Wardi (d. 749 AH). It is known that the grandfather drew a map for the world. He drew it as round, surrounded by the aforementioned mountain.

By the way, I am not of those who could prove today the existence of this mountain around the Earth. However, I could not deny its existence altogether, for I am not qualified to do so. However, for those exegetes who say that had it been existent, people would have been able to see it, I would say to them, "The Earth is extended so far to an extent known only by Allah. He who claims to have reached the end of the Earth is either a liar or delusional. If something is so far, humans cannot look at it even by means of modern technology."

However, the purpose of my citations is to say to Dr. Adnan Ibrahim that there were Muslim scholars who supported the theory that the Earth is flat, and it is not as you hint in your last speech.

However, he who looks into some of the Iranian heritage will find that the Earth is surrounded by a mountainous dam and behind the dam lie devils or beasts. Most young men today do not know this fact, otherwise they would have said that they knew the whereabouts of Gog and Magog and would have linked Cyrus to Dhul-Qarnayn and put forward new hypotheses.

To get back to what I was saying, the well-known geographer Abu Abdullah Zakariyyah Al-Qazwini (d. 682 AH) supported the theory of the flat Earth as manifested in his book *'Aja'ib Al-Makhluqat wa Ghara'ib Al-Mawjudat*. He also held that the Earth is surrounded by a mountainous wall and that there is cosmic water under the Earth, which is held up by the sea monster Bahamoot. However,

I do not mean to prove these wonders stated by Al-Qazwini. My purpose is to prove to Dr. Adnan that some geographers, astronomers, mathematicians, scholars of Sharia, and Sufis held that the Earth is flat. Similarly, brothers Ali and Muhammad Ahmad bin Muhammad Al-Sharfi Al-Safaqsi thought the Earth is flat with a round shape. Ali was a geographer, mathematician, and astronomer who died in 958 AH. Muhammad was also a geographer and botanist. Ali is the author of the book entitled *Al-Tablah*, which included maps of the Earth being surrounded by water and the water being surrounded by mountain ranges surrounding the whole Earth. This supports the new theory of the flat Earth and contradicts the idea of the spherical Earth that floats in the space as an atom.

In his speech, Dr. Adnan made statements to the effect that the whole Muslim nation agrees that the Earth is spherical, but this is not true. I mentioned the opinions of some of the scholars of Ahl Al-Sunnah. Doesn't Dr. Adnan Ibrahim know that the Shiite Scholar Sheikh Al-Mufid Muhammad ibn Muhammad ibn Al-Nu'man Al-'Akbari (d. 413 A.H.), author of *Al-Irshad fi Ma'rifat Hijaj Allah 'Ala Al-'Ibad*, used to say that the Earth is flat and rejected those who hold that it is spherical under the ground that their statement contradicts the Quran, Sunnah, and the Shiite heritage! Doesn't Dr. Adnan himself stated that the Sufi Ibn Al-'Arabi is one of those who hold that the Earth is a flat surface. All praise be to Allah that Dr. Adnan Ibrahim proved that in his speech as well that the exegete Ibn 'Ajibah who was in favor of the theory that the Earth is flat.

Even among the Ibadi sect, there is disagreement. Some of them argue that the Heaven surrounds the Earth or that it is flat, and the edges of the Heaven touch the edges of the Earth (i.e., it is like a dome). Didn't Dr. Adnan, while quoting Ibn Hazm's statement on this issue, say ironically that those who hold that the Earth is flat are the laymen, whereas the well-known scholars hold that it is spherical? First, before explaining Ibn Hazm's statement in detail, aren't laymen among Muslims? From where did the laymen deduce the Earth is flat? Did they not adopt this from the Quran and other sheikhs and scholars? They are Muqallids (following the teachings of a certain school of jurisprudence) in most cases according to you, aren't they? So, tell me, for Allah's sake, why do the laymen hold that the Earth is flat and contradict the well-known scholars? That is weird!

This is a sufficient refutation of the claim that there is agreement on this issue. Ironically, Dr. Adnan Ibrahim himself made us understand that the apparent meaning of the Quranic verses indicates that the Earth is flat, and that there can be reconciliation between them and the modern scientific theories. However, my question to Dr. Adnan is, "Why do you reconcile between the apparent meaning of the Quran and something else?" There is no need for reconciliation unless there is contradiction. Isn't that what we learned from the principles and branches of jurisprudence, my dear professor? Doesn't that mean that if man does not have background information that the Earth is spherical, he would understand from the Quran that the Earth is flat? Isn't that sufficient to refute the argument of those who hold that the apparent meaning of the Quran does not indicate that the Earth is flat? No, it proves so. Otherwise, there would have been no need to make reconciliation. As many scholars of Hadith attempted to reconcile between the weak Hadiths and the Quran, some scholars also tried to reconcile between the output of the scientific community and the Quran without wondering one day "what if the Hadith is weak? Or what if the theory is wrong?"

It is easy to make reconciliation, but it is difficult for the other party to prove that he is right. I have written an article some years ago renouncing the trend to make reconciliation between everything and the Quran, even if that thing for which reconciliation is made is false. This is like what the Christians do when they try to reconcile the Trinity and the Quran in a desperate attempt to prove that it is stated in the Quran.

However, let us examine the claim made by Dr. Adnan, which I found to be inadequate. He claimed that Ibn Taymiyyah reported the agreement of scholars that the Earth is spherical, according to the modern society's understanding of the shape and position of the Earth. However, I think Dr. Adnan Ibrahim did not carry out personal search during his search on the issue. I think he had just read some articles on the internet or some books of contemporary thinkers who were affected by modern scientific theories, or he might read some writings of some enthusiastic youth eager to prove that it was Muslims who first said that the Earth is spherical (according to our understanding today), before other people. However, there is nothing wrong with this at all. I just referred to one of the causes for which Dr. Adnan is to blame. Anyone who notices the order Dr. Adnan followed in his speech and the scattered and repeated articles on the internet would realize why and for what Dr. Adnan is to blame. This is only what I think, and therefore I call upon researchers not to depend only on web articles that are mostly taken out of their context, as many people do so to serve their purposes.

Ibn Taymiyyah proved that there is unanimity of opinion that the heaven is spherical. He said, and I quote, "Heavens are spherical according to Muslim scholars, as more than one of Muslim scholars reported Muslim consensus, including, Abu Al-Husayn Ahmad ibn Ja'far ibn Al-Munadi, a great Hadith scholar and one of the companions of Imam Ahmad, and author of about 400 Hadith compilations. This consensus was reported by Imam Abu Muhammad ibn Hazm and Abu Al-Faraj ibn Al-Juzi. Scholars reported this consensus through well-known chains of narrations from the Companions of the Prophet and Followers. They supported their opinions by proofs from the Quran and Sunnah. They simplified their arguments by traditional textual proofs as well as by mathematical proofs. I do not know that any known Muslim scholar denied that, unless a few sects of dialectical scholars, who, when debating with astrologers, said by way of possibility that the heaven may be square or hexagonal, but did not deny that it may be round and permitted otherwise. I do not know that anyone said, and affirmed, that it is spherical, unless he is an ignorant person."

In another position, he said, "They have also unanimously agreed that the Earth, with all its movements in land and sea, is similar to a sphere." He said: "This is evidenced by the fact that the sun, moon, and planets do not rise and set concurrently on all parts of the Earth."

Ibn Taymiyyah proved first that there is a consensus that the Heaven is spherical. But even if he said there is a consensus that the Heaven and Earth are spherical, this does not contradict what I maintained in my previous article, for the Heavens and Earth are a set that may be like a sphere, egg, or bubble. This contradicts neither the Quran, the Old Testament, nor what was customarily thought in the past. If you admitted that the Heavens are like a dome over the surface of the Earth, and the surface of the Earth is flat while what is under the earth is circular, this would not prove that only the Earth is spherical, but rather that the Heavens and Earth are spherical. This is the opinion maintained by Ibn Taymiyyah. It is not as Adnan Ibrahim claimed that there is a consensus that the Earth, and not the Heaven, is spherical.

However, even if we assume that he has already meant that the Earth, and not the Heaven, is spherical, or that both of them are spherical, as maintained by the ancient Greek philosophers, his speech gave this meaning, unless Dr. Adnan did not understand that. My question is, "Who said what is meant is that it resembles the ball we know today, and is not spherical and flat, that is, with a flat surface in a circular shape?" This is also possible and does not contradict the fact that the Earth is flat and also does not correspond with the view that the Earth is spherical as we know today. So, where is that alleged consensus?

Does Ibn Taymiyyah not quote as evidence in the same Fatwa the transmitted narration that reads, "The Heaven is a dome over the Earth." He quoted Iyas ibn Mu'awiyah as saying, "The Heaven is like a dome over the Earth." This makes sense, for if the Earth is a flat circle and the Heaven is spherical and round, it is like the dome. He then quoted the statement of Ibn Al-Munadi, may Allah be merciful to him, which Adnan used as evidence that the Earth is spherical, without saying that the Earth is a point in the realm of the Heaven. However, it is likely understood from his speech that the Heavens surround the Earth and that its underneath is in the ultimate bottom. He said, and I quote, "He who knows that orbits are circular and the circumference, which is the ceiling, is the 'Illiyyin (the very highest levels of Heaven) and the center, which is the inner bottom of the earth, is Sijjin (the very bottom of hell) and the lowest of the low, would come to understand from Allah's contrast between the 'Illiyyin and the Sijjin that the contrast is apparently between the high and the low or between the vast and narrow, because the high entails vastness and the narrow entails lowness. As such, he would come to know that the Heaven is absolutely above the Earth and could never be imagined to be under it, even if it is circular and surrounds it."

However, the purport of the speech of Ibn Al-Munadi is different from what Dr. Adnan Ibrahim maintained. I think his speech is likely similar to the speech of ancient philosophers who perceived all orbits as circular in shape, with the

Earth circular in the center and around which other orbits rotate. However, I may be wrong. Allah knows better.

Anyone who studied and noticed the maps drawn by Muslim scholars, who are reported to have said that the Earth is spherical, would find them draw the Earth as a circle. This supports some of the orientations of those who adopt the theory that the Earth is flat and who identify the place of Gog and Magog and the disk surrounding the Earth in the north. You can find this in more than one map. You can google them, and you will find what I mean.

However, Ibn Hazm uses the following verse as evidence, **"He wraps the night over the day and wraps the day over the night"** [Al-Zumar:5]. Before commenting on the interpretation of the verse, I will first say, "Who said that Ibn Hazm meant the ball we know today? How will you prove that he did not mean it is flat and circular in shape? Didn't he say soon later that there is nothing under the Earth?" He likened it to the act of wrapping a turban around one's head and said it is like "winding it." In fact, a turban is worn by winding and wrapping it around the center, as when an Omani or Emirati person does when he wears his turban. Perhaps this is what the late scholars understood from the speech of Ibn Hazm. Those late scholars, who tend to hold the belief that the Earth is spherical, as we perceive the ball today, did not say that Ibn Hazm said it is so. He who reads these pages by Ibn Hazm will notice that he supports the idea that the sun and the moon move according to the new concept of the flat Earth theory regarding the movement of the sun and moon.

The third possible meaning of Ibn Hazm's speech is that the Earth is spherical from the top and flat from the bottom, and under it are water, air, and the like. It is like the egg, with the yolk flat from its bottom and circular from its top. It is sometimes said that the 'Arsh (Allah's Throne) surrounds the Heaven and the Earth just like an egg. That is why the late scholars thought that this referred to the ball we know today. In fact, it is meant that the Earth is the center and in the middle of the universe, and some of the ancient philosophers regarded the center as being flat, while others regarded it as spherical. Many astronomers said that the Heavens and the 'Arsh surround the Earth and the Earth is like a sphere inside an egg. They said, for example, as Ibn Khordadbeh said: "The Earth is as round as the ball." In another position he said, "It is like the yolk inside an egg." How does the water and the yolk inside the egg look? It is flat from the bottom and spherical or round from the top (this is if there is vacuum, but if water surrounds everything, their speech is true. This brings us to the theory of the Earth as the center of the universe surrounding the Heavens). However, the surrounding crust, according to them, is the 'Arsh or the Heavens. This is acceptable and does not contradict the idea of the flat Earth and doesn't support the idea that the Earth is spherical as much as it supports the idea of the flat Earth. Consider and test that with respect to the egg, and you will find my speech, Allah willing, correct.

Let us get back to Ibn Hazm's consensus, on account of which Dr. Adnan Ibrahim, may Allah bless him, claimed that there is unanimity on the spherical-ness of the Earth, with the present shape. This is the beginning of his speech: "No one of the Muslim Imams and scholars, may Allah be pleased with them, denied that the Earth is spherical. No argument is reported for any of them in this respect. Demonstrations from the Quran and Sunnah supported that the Earth is spherical." He cited a group of proofs in this respect in his book, *Al-Fasl fi Al-Milal wa Al-Ahwa' wa Al-Nihal* (2/78).

However, Ibn Hazm did not mean that the Earth is spherical in the sense we understand today. His speech instead refers to the flat Earth and the Heavenly dome. Ironically, the arguments of the opponent turn to be proofs in favor of the side holding the true view. After reviewing Ibn Hazm's speech in this chapter, you will find him refute the objection of Judge Mundhir ibn Sa'id who regarded orbits different from the Heavens. The Judge believed that the Heavens are above the Earth, for had they been surrounding the Earth, some Heavens will be under the Earth.

This gives us evidence that the Judge did not consider the Earth as spherical as we understand today; he instead considered it as flat. Ibn Hazm said in his refutation, "This is nothing, for the thing which is described as under something will be above another thing, except for the center of the Earth, for it is in the very bottom and has nothing under it." Then he said, after explaining that there is nothing above the top of the orbit, "The Earth, according to this evidence, is necessarily the place under the Heavens. As such, the Heaven is above the Earth and, by contrast, the Earth is under the Heaven. Any human, wherever he is, will have his head toward the Heaven and his legs toward the Earth." This speech does not correspond with the concept of the spherical Earth that floats in the space, as we understand today, but rather corresponds completely with the Earth being the center of the universe, We can say that Ibn Hazm meant that the small sphere of Earth, which is in the Heaven, is at the very bottom. It is not seen moving, and the sun cannot be seen under it, for it does not depart from the Earth, according to him, and, thus, this is related to the issue of the circular disc.

Ibn Hazm also pointed out, "The sun and the moon cannot part from, or get out of the Heavens, for however they rotate, they are in the Heavens. As such, the Heavens are layers, one above the other over the Earth." He then narrated the Hadiths that supported his speech, such as, **"The Heaven is like a dome over the Earth"** and, **"His Throne is above His Heavens and the Earth as such."** He then reported the Hadiths that indicate that the Paradise is above the Heavens and under the 'Arsh, and the Hell is under the Earth. Accordingly, we, my fellows, have come to know that Ibn Hazm meant that his opinion corresponds with the Quranic verses and the prophetic Hadith, namely that the Heavens are layers in the top and the Earth is in the bottom. His speech supports the new, popular perspective of the flat Earth, held by many Westerns, where the Arctic is in the middle of the Earth.

To get back to Dr. Adnan's and other contemporary scholars' evidence based on Allah's Saying, **"He wraps the night over the day and wraps the day over the night"** [Al-Zumar:5]. First, what is the relation between His Saying **"He wraps the night over the day"** and the Earth being spherical. There is no relation at all. None of the early exegetes said this verse indicates that the Earth is spherical. Anyone who makes such a claim is a liar. Adnan erred, like other contemporary scholars, when he cited this verse as evidence. Refer to the interpretation of this verse, and you will find Al-Tabari says, "This covers that and that covers this, as Allah says, **'He causes the night to enter the day, and He causes the day to enter the night'** [Fatir:13]." Ibn 'Abbas said, "He brings the night over the day." In another narration, "The decrease in the hours of the night is added to the hours of the day, and the decrease in the hours of the day is added to the hours of the night." Mujahid said, "Passes into it." Qatadah said, "This covers that, and that covers this." Al-Suddi said, "He causes the day to come and the night to go, and then causes the night to come and the day to go." Ibn Kathir said, "He has subjugated them and He causes them to alternate without ceasing, each seeking the other rapidly, as He says: '**He covers the night with the day, [another night] chasing it rapidly**'" [Al-A'raf:54]. Al-Baghwi, Al-Hasan, Al-Kalbi, and other exegetes held similar opinions. But the contemporary scholars who were affected by the new knowledge used the verse as an evidence only for the opinion they chose to prove that "the Quran gave prior information that were later evidenced by humanity," as Ibn 'Ashur put it. In fact, the idea of the spherical Earth is existent before the revelation of Quran and its being recited by humans. This is, of course, known for those interested in scientific miracles of the Quran. Dr. Adnan Ibrahim also hinted to this fact, may Allah reward him well.

Al-Zamakhshari, who is known as one of the Mu'tazilah, said, "Takwir means wrapping up and winding, just like a turban is wrapped up and wound around one's head. This indicates that the night and the day run in succession, when this vanishes, the other comes, covers and wraps it. It also indicates that each one of them goes away with the other, and this resembles wrapping it up so that it is out of sight. It also indicates that each one, just like a turban, is folded and wound round and round."

However, Al-Fakhr Al-Razi, who argues that the Earth is spherical, says in his interpretation of this verse, "This verse '**He wraps the night over the day and wraps the day over the night**' [Al-Zumar:5] refers to the different cases of the night and day. The light and darkness are two great agents, and each one overcomes the other every day. This indicates that each one of them is overcome and subjected, and so there should be a conqueror and subjugator of both of them, whom they are under his dominion and control, which is Allah, Exalted and Glorified is He." This *Takwir* means that the decrease in the hours of one of them is added to the hours of the other. *Takwir* regarding the night and day means the sense reported in the Hadith: "*We seek refuge with Allah from Al-Hawr, i.e. decrease or recession, after Al-Kawr, i.e. increase or accession.*" This meaning is expressed by Allah, Exalted and Glorified is He, in His Saying, "**He wraps the night over the day and wraps the day over the night**" [Al-Zumar:5]; and His Saying, "**He covers the night with the day**"

[Al-A'raf:54]; and His Saying. **"He causes the night to enter the day, and He causes the day to enter the night"** [Fatir:13]; as well as His Saying. **"And it is He who has made the night and the day in succession for whoever desires to remember"** [Al-Furqan:62]. Had this verse been a textual proof of the sphericalness of the Earth, why didn't Al-Razi, who is referred to as one of those supporting the theory of the spherical Earth, use it as a proof? I just want to say to my dear Dr. Adnan and other contemporary scholars as they said to the opponents of their theory, "Are you more knowledgeable than Al-Fakhr Al-Razi and the great exegetes?"

In Tafsir Al-Samarqandi concerning Allah's Saying, **'He wraps the night over the day,'** Mujahid said, "It means He winds the night around the day," while **"and wraps the day over the night"** means "He winds the day around the night." Muqatil said, "It means that one encroaches upon the other, i.e. they take from the length of one another." Al-Kalbi said, "He takes from the length of the day and adds it to the night such that the night is longer than the day, and then takes from the length of the night and adds it to the day such that the day is longer than the night." Al-Qatbi said, "This goes into that." *Takwir* originally means wrapping and winding, as He says, **"When the sun is wrapped up [in darkness]"** [Al-Takwir:1].

However, Al-Mawardi, who usually reports the different opinions about the interpretation of a Quranic verse, said, "His Saying **'He wraps the night over the day and wraps the day over the night'** [Al-Zumar:5] has three interpretations: First, He causes the night to go into the day and causes the day to go into the night, as maintained by Ibn Abbas. Second, the darkness goes away with the light, and the light goes away with the darkness, according to Ibn Qatadah. Third, one takes from the length of the other, so when the night is short, the day becomes longer, and when the day is short, the night becomes longer, according to Ibn Al-Dahhak. It may probably have a fourth meaning, namely, that He gathers the night until the day spreads out, and then gathers the day until the night spreads out." However, Al-Mawardi did not mention anything that has to do with the theory of the sphericalness of the Earth, nor did he interpret Takwir as scientists today do. There is neither might nor power except with Allah.

On the other side, Al-ʿIzz ibn Abdul-Salam said, **"He wraps the night over the day and wraps the day over the night"** [Al-Zumar:5], which means each one goes into the other, or the night covers the day and goes away with its light and the day covers the night and goes away with its darkness, or that the shortness of each one of them is added to the length of the other.

However, *Tafsir Abu Al-Suʿud*, which Dr. Adnan cited as evidence, has in fact stated something that proves the movement of the Heaven as in the theory of the Heavenly Dome. He said, "It is clarification of the way through which Allah handles them after clarifying their creation. The existence of night and day on Earth is subject to the movement in the Heaven. Each one covers the other or wraps it as clothes are wrapped around a person or winds around the other as a turban wound around the head."

However, Ibn ʿArafah said, "The verse indicates that the Heaven is spherical, for the Takwir or wrapping of the day and night entails necessarily that their place is spherical, for it is impossible that they have no place."

Many exegetes make statements to the same effect, so from where did you conclude that there is a consensus that the verse indicates the Earth is spherical? Where is the evidence? It is only that which the contemporary scholars said and was adopted by Dr. Adnan Ibrahim and others, who claimed that Takwir entails that the Earth is spherical. What a grave error! If you were to imagine the movement of the night and day on the flat Earth, as known in the symbols of Yin and Yang, you would find this closer to the meaning of "wrapping the night over the day" and the winding of a turban. If you considered the Heaven spherical and that it is the place of the day and night, you would find this in accordance with the verse as maintained by Ibn ʿArafah. If you were to combine the two meanings, you would find them in accordance with the verse as well. Why claim without knowledge that the verse indicates only that the Earth is spherical, though none of the early exegetes said that. Even Al-Razi did not say that.

Besides, what do you say about His Saying, **"When the sun is wrapped up [in darkness]"** [Al-Takwir:1]? Didn't most exegetes interpret it as meaning that the sun lost its light or that it is folded up until its light is lost? Or that the sun now is not spherical, and Allah will make it spherical on the Day of Judgment? Moreover, who said that the night and day are just incidental circumstances and have no material reality, and it is a matter of only sunrise and sunset? Who said that? Is there consensus on such a claim as well? What if Allah created the night and day before the Heavens and the Earth? Did you think of that before, Dr. Adnan? Why cannot it be so? Do you not see that Allah mentions night and day before sun and moon in many Quranic verses? In the beginning of Chapter of Al-Zumar, it seems as if there is order of creation as Allah, Exalted and Glorified is He, says, "**The revelation of the Quran is from Allah, the Exalted in Might, the Wise. Indeed, we have sent down to you the Book, [O Muhammad], in truth. So worship Allah, [being] sincere to Him in religion. * Unquestionably, for Allah is the pure religion. And those who take protectors besides Him [say], "We only worship them that they may bring us nearer to Allah in position." Indeed, Allah will judge between them concerning that over which they differ. Indeed, Allah does not guide he who is a liar and [confirmed] disbeliever. * If Allah had intended to take a son, He could have chosen from what He creates whatever He willed. Exalted is He; He is Allah, the One, the Prevailing. * He created the heavens and earth in truth. He wraps the night over the day and wraps the day over the night and has subjected the sun and the moon, each running [its course] for a specified term. Unquestionably, He is the Exalted in Might, the Perpetual Forgiver. * He created you from one soul. Then He made from it its mate, and He produced for you from the grazing livestock eight mates. He creates you in the wombs of your mothers, creation after creation, within three darknesses. That is Allah, your Lord; to Him belongs dominion. There is no deity except Him, so how are you averted?"** [Al-Zumar:1–6]. Didn't Allah, Exalted and Glorified is He, say, "**And it is He who created the night and the day and the sun and the moon; all [heavenly bodies] in an orbit are swimming**" [Al-Anbiya':33]? Didn't Allah say, "**Say, 'Have you considered: if Allah should make for you the night continuous until the Day of Resurrection, what deity other than Allah could bring you light? Then will you not hear?' * Say, 'Have you considered: if Allah should make for you the day continuous until the Day of Resurrection,** what deity other than Allah could bring you a night in which you may rest? Then will you not see?' * And out of His mercy He made for you the night and the day that you may rest therein and [by day] seek from His bounty and [that] perhaps you will be grateful" [Al-Qasas:71–73]? Didn't Allah also say, "And of His signs are the night and day and the sun and moon" [Fussilat:37]?

However, I do not deny that there is a relationship between the sun and moon and the night and day. However, this should not make us interpret the verses as such fellows did without considering the Book of Allah and contemplating its meaning well. Also, notice Allah's Saying, "And it is He who created the night and the day and the sun and the moon; all [heavenly bodies] in an orbit are swimming" [Al-Anbiya':33]. You will see that the Earth is not added here or described as swimming in an orbit. This is worthy of consideration.

Surprisingly, I found in the Torah or the Old Testament the same matter I found in the Quran, namely that Allah created night and day before sun and moon. As I talked about the Torah, Dr. Adnan Ibrahim stated that the Torah or the Old Testament indicates the Earth is flat. This is what I mentioned in my article, *Preference for Truth over the Issue of Creation*, that the Torah, as well as the Quran, states that the Earth is flat. Unfortunately, the Christians and Muslims, out of ignorance, ridicule each other, accusing each other of being against knowledge. Dr. Adnan and others claimed that the Torah is against knowledge and that the Torah states that the Earth is flat. Similarly, the Christians challenge the Quran because it states that the Earth is flat. Unfortunately, neither the Muslims nor the Christians questioned the theory of the spherical Earth.

However, if Dr. Adnan and others say that the Torah is distorted, I ask them to give evidence, which will be difficult to them. However, let us assume that it is distorted. Why didn't Allah then deny anything mentioned in the Torah regarding the creation of Heavens and Earth, but rather they—the Quran and Torah—are seemingly identical and in agreement on this point. This is contrary to his statement about the Lord's rest in the Old Testament; there was a misunderstanding, so Allah said in the Quran, "And We did certainly create the heavens and earth and what is between them in six days, and there touched

Us no weariness" [Qaf:37]. Also, regarding Sulayman, Allah said, **"It was not Solomon who disbelieved"** [Al-Baqarah:102]. Also, regarding Jesus, Allah denied Trinity and that he is the son of Allah. Why didn't the Quran deny the information reported in the Torah or the Old Testament about the creation and shape of the Heavens and Earth? Allah said the truth when He said, **"Say, 'Then bring a scripture from Allah which is more guiding than either of them that I may follow it, if you should be truthful'"** [Al-Qasas:49].

Dr. Adnan Ibrahim argues that Allah's Saying, **"And the earth—We have spread it"** [Al-Hijr:19], and His Saying, **"And it is He who spread the earth"** [Al-Ra'd:3], indicates that the Earth is spherical, because it is only the spherical bodies that can be spreadable in any place you are standing on. Such an assumption is not necessarily the likely meaning of the verse. It is true that Al-Razi refuted the question by saying that the Earth is spherical, not because it is always spread out but rather because of its large size, man always sees it spread out. However, I refuted all these assumptions in my article, *Preference for Truth over the Issue of Creation.*

Let us consider the opinions of the exegetes.

Allah, Exalted and Glorified is He, says, **"And the earth—We have spread it and cast therein firmly set mountains and caused to grow therein [something] of every well-balanced thing"** [Al-Hijr:19]. Al-Tabari said: "His Saying **'the earth—We have spread it'** means that He has stretched it out and spread it flat, and His Saying **'and cast therein firmly set mountains'** means that He placed firm mountains on Earth as Bishr reported from Yazid from Sa'id from Qatadah. In another verse, Allah says, **'And after that He spread the earth'** [Al-Nazi'at:30]. It is reported that the Earth was spread out from Umm Al-Qura, Mecca. Al-Qurtubi said, Ibn 'Abbas interpreted it as meaning, 'We spread out the Earth on water' as He said elsewhere, **'And after that He spread the earth,'** meaning spread it flat. And, **'And the earth We have spread out, and excellent is the preparer'** [Al-Dhariyat:48]. This provides refutation to those who claim that it is like a sphere."

Ibn Kathir said, "He created the Earth, stretched it and spread it flat." Most exegetes hold the opinion that He has spread it on water. In *Tafsir Al-Khazin*, His

Saying **"the earth—We have spread it"** means we have spread it on water. It is said that it was spread out from underneath the Ka'bah, as held by many Exegetes. Astronomers claim that it is a great ball, with part of it in water and the other part outside water. As for Allah's Saying **"the earth—We have spread it,"** they argued that if the ball is so great, each part of it would be like a great flat area, and, thus, it is established that the Earth is stretched out and spread flat and is a ball. This argument was rejected by exegetes because Allah narrated in His Book that the Earth is stretched out and spread flat, and had it been a ball, He would have said so. Allah knows better how the Earth is stretched.

After this explanation, we find that none of the early exegetes adopted the verse as evidence that stretching refers only to the spherical Earth. This view is adopted only by the late exegetes, including Sheikh Al-Sha'rawi. If we are to consider the verse as it is, why it is not possible that the Earth is stretched to what is known only by Allah. If one asked where the edge of the Earth is, I would answer him that no one reached its edge. Besides, who knows if it has edges that people can discern. Why couldn't it be stretched out to an end known only by Allah? What about Allah's Saying, **"And the heaven We constructed with strength, and indeed, We are [its] expander"** [Al-Dhariyat:47]? The Earth is stretched magically, if you accept the theory of the Heavenly dome, which is in accordance with the rest of verses and the Torah, as Allah stretched out the Earth on the water that is under the Earth. Most exegetes said that the verse can be interpreted as meaning that Allah spread out the Earth, and it is in accordance to what is stated in the Torah. As everlastingness does not entail eternity, stretching does not entail endlessness. Take this into consideration, and the evidence is that Allah will stretch the Earth on the Day of Judgment as well, as He says, **"And when the earth has been extended"** [Al-Inshiqaq:3].

The problem is that those people think that the Earth is a planet floating in the space, though we do not have a Quranic verse, Hadith, or a statement by a Companion or Follower that indicates the Earth is a planet.

Most of the scholars who claim the Earth is spherical are influenced by the books of astronomy and philosophy, and some of them—such as Al-Fakhr Al-Razi—even wrote books on this subject. What is expected from a man who wrote on this subject and believed it to be true from the beginning when he interpreted the Quran? He would surely try to reconcile between the Quranic verses and the sciences he learned. Likewise, those scholars affected by the traditions attributed to the Prophet, peace be upon him, and the Companions, try frequently to reconcile between the Quranic verses and the traditions, even if these traditions are not correct, only because they think them true. This can be noted by anyone who makes serious studies on this subject.

Al-Fakhr Al-Razi was much influenced by philosophy and astronomy until he became known for this among scholars and the elites. You do not need to read what was said about him or his books. It is well known, and I do not have to explain it more here.

Allah, Exalted and Glorified is He, said, **"From the earth We created you, and into it We will return you, and from it We will extract you another time"** [Taha:55]. This explicitly indicates that human will remain on this Earth. Allah also said, **"Say, 'Travel through the land'"** [Al-An'am:11], and did not say "travel or fly in the Heavens." He did not say so, not only because the human mind would not understand it at that time or that was not the purpose, but also because they could not do so.

All that people can do with the Heaven is to look at it, consider it and fly in the atmosphere of the sky, at best, which I will talk about later. Allah, Glorified is He, says, **"Say, 'Observe what is in the heavens and earth.' But of no avail will be signs or warners to a people who do not believe"** [Yunus:101]; and, **"Do they not look into the realm of the heavens and the earth and everything that Allah has created and [think] that perhaps their appointed time has come near? So in what statement hereafter will they believe?"** [Al-A'raf:185]; and, **"and give thought to the creation of the heavens and the earth,"** [Al 'Imran:191]. When the Earth is mentioned without the Heavens, Allah says, **"Travel through the land"** [Al-An'am:11]. **"It is He who made the earth tame for you—so walk among its slopes"** [Al-Mulk:15]. This suggests a certain meaning that you have to consider.

It is true that the sphericalness of the Earth is an ancient idea. I do not deny this, and I am not among those who claim that there is a Great Conspiracy

Theory, though it is not unlikely that there may be a conspiracy or low-level conspiracies. However, while the idea of the spherical Earth is old, the view of the flat Earth is older. The view that the Earth was flat was adopted by a number of well-known Pre-Socratic philosophers, including Thales, Democritus, and Anaximander, who maintained that the Earth is a cylinder whose top and bottom are flat (many of those holding that the Earth is flat follow probably his view). Also, Anaximenes, his disciple, maintained that the Earth is a flat disc. The same was adopted by philosophers such as Leucippus and Pyrrho. Also, Thales assumed the Earth is a floating piece of wood over water. This is very close to the perspective I presented in my article *Preference for Truth over the Issue of Creation*. This view explains why the Earth used to sway until Allah fixed it with mountains. Many world and religious legends say the Earth is flat, and the Heavens are like a dome or ceiling above it.

I would like to note that some ancient scholars who said that the Earth is spherical also said that the substance of water is spherical. They said that spherical water surrounds the spherical Earth and the lord or god (according to the doctrinal background) brought out a part of the sphere of Earth from the sphere of water, and thus the top and bottom parts are water. However, this does not also support the theory of the spherical Earth that floats in the space as an atom, as compared by the sun. This rather proves the other theories that relate to the flat Earth. It is not proper to accept the statement of the old philosophers that the Earth is spherical without considering their general view of the creation. This makes one fall into the error into which the contemporary scholars, including Dr. Adnan Ibrahim, fell.

However, I reviewed all what Dr. Adnan said and found that he did not, in fact, offer one refutation of all the proofs I stated in my previous article and quoted from the Holy Quran. He interpreted neither the direction of the Qibla, nor the deluge of Noah, nor the movement of the mountains. He who reads my aforementioned article will find a strong perspective in accordance with the Holy Quran, Torah, Hadith, legends, religions, and the statements of some philosophers, scholars, and thinkers. It is a comprehensive perspective that explains many matters without contradiction or affected interpretation. All praise is due to Allah.

Pathetically, some fellows attempt to interpret the Quranic verses to accord with the scientific theories they follow. Dr. Adnan Ibrahim reviled those holding the Earth is flat and claimed more than one time that he did not find any exegetes who support their views. Nevertheless, I, with my limited knowledge, found what he could not find. I say to Dr. Adnan Ibrahim, "You came to know some things and failed to know others. You should have searched this topic seriously and objectively, and not just for looking for articles to refute the idea of the flat Earth."

I advise my dear fellow, Dr. Adnan, whom I really love, and I advise myself, to reconsider this Quranic verse, **"And do not pursue that of which you have no knowledge"** [Al-Isra':36]. I remind myself and anyone who might think to refute or ridicule Dr. Adnan to recite this Quranic verse that contains great lessons, **"[Moses] said, 'My Lord, forgive me and my brother and admit us into Your mercy, for You are the most merciful of the merciful'"** [Al-A'raf:151].

Peace and blessings of Allah be upon you.